D1034027

SPECTRAL THEORY OF ORDINARY DIFFERENTIAL OPERATORS

ELLIS HORWOOD SERIES IN MATHEMATICS AND ITS APPLICATIONS

Series Editor: Professor G. M. BELL, Chelsea College, University of London

The works in this series will survey recent research, and introduce new areas and up-to-date mathematical methods. Undergraduate texts on established topics will stimulate student interest by including present-day applications, and the series can also include selected volumes of lecture notes on important topics which need quick and early publication.

In all three ways it is hoped to render a valuable service to those who learn, teach, develop and use mathematics.

MATHEMATICAL THEORY OF WAVE MOTION
G. R. BALDOCK and T. BRIDGEMAN, University of Liverpool.
MATHEMATICAL MODELS IN SOCIAL MANAGEMENT AND LIFE SCIENCES
D. N. BURGHES and A. D. WOOD, Cranfield Institute of Technology.
MODERN INTRODUCTION TO CLASSICAL MECHANICS AND CONTROL
D. N. BURGHES, Cranfield Institute of Technology and A. DOWNS, Sheffield University.
CONTROL AND OPTIMAL CONTROL
D. N. BURGHES, Cranfield Institute of Technology and A. GRAHAM, The Open University, Milton Keynes.
TEXTBOOK OF DYNAMICS
F. CHORLTON, University of Aston, Birmingham.
VECTOR AND TENSOR METHODS
F. CHORLTON, University of Aston, Birmingham.
TECHNIQUES IN OPERATIONAL RESEARCH
VOLUME 1: QUEUEING SYSTEMS
VOLUME 2: MODELS, SEARCH, RANDOMIZATION
B. CONOLLY, Chelsea College, University of London.
MATHEMATICS FOR THE BIOSCIENCES
G. EASON, C. W. COLES, G. GETTINBY, University of Strathclyde.
HANDBOOK OF HYPERGEOMETRIC INTEGRALS: Theory, Applications, Tables, Computer Programs
H. EXTON, The Polytechnic, Preston.
MULTIPLE HYPERGEOMETRIC FUNCTIONS
H. EXTON, The Polytechnic, Preston
COMPUTATIONAL GEOMETRY FOR DESIGN AND MANUFACTURE
I. D. FAUX and M. J. PRATT, Cranfield Institute of Technology.
APPLIED LINEAR ALGEBRA
R. J. GOULT, Cranfield Institute of Technology.
MATRIX THEORY AND APPLICATIONS FOR ENGINEERS AND MATHEMATICIANS
A. GRAHAM, The Open University, Milton Keynes.
APPLIED FUNCTIONAL ANALYSIS
D. H. GRIFFEL, University of Bristol.
GENERALISED FUNCTIONS: Theory, Applications
R. F. HOSKINS, Cranfield Institute of Technology.
MECHANICS OF CONTINUOUS MEDIA
S. C. HUNTER, University of Sheffield.
GAME THEORY: Mathematical Models of Conflict
A. J. JONES, Royal Holloway College, University of London.
USING COMPUTERS
B. L. MEEK and S. FAIRTHORNE, Queen Elizabeth College, University of London.
SPECTRAL THEORY OF ORDINARY DIFFERENTIAL OPERATORS
E. MULLER-PFEIFFER, Technical High School, Erfurt.
SIMULATION CONCEPTS IN MATHEMATICAL MODELLING
F. OLIVEIRA-PINTO, Chelsea College, University of London.
ENVIRONMENTAL AERODYNAMICS
R. S. SCORER, Imperial College of Science and Technology, University of London.
APPLIED STATISTICAL TECHNIQUES
K. D. C. STOODLEY, T. LEWIS and C. L. S. STAINTON, University of Bradford.
LIQUIDS AND THEIR PROPERTIES: A Molecular and Macroscopic Treatise with Applications
H. N. V. TEMPERLEY, University College of Swansea, University of Wales and D. H. TREVENA, University of Wales, Aberystwyth.
GRAPH THEORY AND APPLICATIONS
H. N. V. TEMPERLEY, University College of Swansea.

SPECTRAL THEORY OF ORDINARY DIFFERENTIAL OPERATORS

ERICH MÜLLER-PFEIFFER, Dr.rer.nat.habil.
Professor of Mathematics
Pädagogischen Hochschule 'Dr. Theodor Neubauer'
Erfurt/Mühlhausen, DDR

Translation Editor:
M. S. P. EASTHAM, M.A., D.Sc.
Chelsea College, University of London

ELLIS HORWOOD LIMITED
Publishers · Chichester
Halsted Press: a division of
JOHN WILEY & SONS
New York · Chichester · Brisbane · Toronto

First published in 1981 by

ELLIS HORWOOD LIMITED
Market Cross House, Cooper Street, Chichester, West Sussex, PO19 1EB, England

The publisher's colophon is reproduced from James Gillison's drawing of the ancient Market Cross, Chichester.

Distributors:

Australia, New Zealand, South-east Asia:
Jacaranda-Wiley Ltd., Jacaranda Press,
JOHN WILEY & SONS INC.,
G.P.O. Box 859, Brisbane, Queensland 40001, Australia

Canada:
JOHN WILEY & SONS CANADA LIMITED
22 Worcester Road, Rexdale, Ontario, Canada.

Europe, Africa:
JOHN WILEY & SONS LIMITED
Baffins Lane, Chichester, West Sussex, England.

North and South America and the rest of the world:
Halsted Press: a division of
JOHN WILEY & SONS
605 Third Avenue, New York, N.Y. 10016, U.S.A.

British Library Cataloguing in Publication Data
Müller-Pfeiffer, Erich
Spectral theory of ordinary differential operators. –
(Ellis Horwood series in mathematics and its applications).
1. Differential operators
2. Spectral theory (Mathematics)
3. Eigenvalues
I. Title II. Spektraleigenschaften singularer gewohnlicher Differentialoperatoren.
English
515.7242 QA329.4 80–42097

ISBN 0–85312–189–3 (Ellis Horwood Limited, Publishers)
ISBN 0–470–27103–5 (Halsted Press)
Printed in Great Britain by R. J. Acford Ltd., Chichester.

Table of Contents

Foreword

Professor Müller-Pfeiffer has worked in the fields of differential geometry and, during the last ten or so years, the spectral theory of differential operators. His research papers are notable for their careful style, penetrating results and, in the choice of topics, their good taste.

The purpose of this translation is to make more generally available not only Professor Müller-Pfeiffer's own methods and results but also his view of the subject as a whole. The book can be read as a sequel to Glazman's well-known work (reference [15]); some of the results in [15] are extended here while other topics, such as the non-existence of eigenvalues in the essential spectrum, are not covered in [15] .

The initial translation into English was provided by the author himself. My task has been to put the text into idiomatic English without unduly imposing my own style, and I wish to acknowledge with gratitude the help of my father, Mr M. Eastham, in clarifying a number of grammatical points. I have also made certain changes in the terminology and notation. More important, Professor Müller-Pfeiffer has provided additional material for this edition including a completely new chapter – Chapter 6 on oscillation criteria – and he has also expanded the bibliography.

M. S. P. EASTHAM
Chelsea College
London University

Author's Preface

In this monograph the spectrum of ordinary self-adjoint differential operators of even order is investigated, where the fundamental differential expression is given by

$$\sum_{k=0}^{n} (-1)^k \frac{\mathrm{d}^k}{\mathrm{d}x^k} a_k(x) \frac{\mathrm{d}^k}{\mathrm{d}x^k}, \quad a_k(x) \text{ real-valued}, a_n(x) > 0, a < x < b.$$

The differential operators are singular, by which we mean that either the basic interval (a,b) is infinite or at least one of the functions $a_n^{-1}(x), a_k(x) (k=0,\ldots,n-1)$ is not summable on (a,b). The basis for the investigation is the theory of linear operators in Hilbert space. The ideas from this theory which are used in this book are summarized in Chapter 1.

We are mainly concerned with three aspects of the analysis of the spectrum: the essential spectrum, the discrete spectrum, and the non-existence of eigenvalues. In Chapter 2, on the basis of theorems by M. S. Birman on the perturbation theory of quadratic forms and with a generalization of one of these theorems, the location of the essential spectrum is described in terms of the behaviour of the coefficients. Also a section deals with the perturbed Euler differential operator. The special case that no essential spectrum exists (a discrete spectrum) is handled in Chapter 3. Here, by means of a theorem by Rellich, we give necessary and sufficient conditions for the discreteness of the spectrum. In Chapter 4 self-adjoint operators are defined by means of semibounded bilinear forms. Both the boundary conditions which appear and the coefficients $a_k(x)$ influence the occurrence of eigenvalues, and criteria are formulated for the case that certain intervals on the real axis contain no eigenvalue. An important method in proving the non-existence of eigenvalues is that of scale transformation. This method is generalized for differential operators of second order in Chapter 5. Moreover, as a new tool, an oscillation theorem for the solutions of Sturm-Liouville equations is used. In this way the results of Chapter 4 can be strengthened for operators of second order. There is an appendix which contains general results further to those in Chapter 1. In order to limit the length of the book, we have omitted certain topics. We do not cover, for example, the asymptotic distribution of eigenvalues in the case of a discrete spectrum and the existence, and estimation of the length of, gaps in the essential spectrum. Nor do we consider extensions to elliptic differential operators of arbitrary order.

In part the monograph contains the results of the author's research on ordinary differential operators during recent years. But some results are new. Certain theorems in Chapter 2 and 3 are generalizations of well-known results that can be found in the monograph *Direct Methods of Qualitative Spectral Analysis of Singular Differential Operators* [15] by I. M. Glazman. The operator $-d^2/dx^2 + q(x)$ (Schrödinger operator), which is the most frequently examined in the literature is a special case in the various chapters. But there are many results about its spectrum which could not be considered here. For such results we refer to the books of N. Dunford and J. T. Schwartz [8] and I. M. Glazman [15] for instance.

I am indebted to Professor H. Triebel (Jena) for suggesting that I write this book. He read the manuscript and made critical comments, and I thank him especially. Further I wish to thank Dr A. Weber for a critical reading of part of the manuscript, and I thank my wife for preparing the typescript.

Chapter 1

Fundamental Concepts

In Chapter 1 as well as in the appendix, those notations and theorems are made available which are essential for the examination of the spectra of differential operators within the context which we have in mind. Thus, those methods which have a direct bearing on operators in Hilbert space are introduced in the first chapter: other methods, such as Embedding Theorems for Sobolev spaces, are listed in the appendix. The fundamental theorems in this chapter and in the appendix will be proved only if the proofs are relatively short or if they differ from those of other authors in the literature. Such theorems as the Spectral Theorem for self-adjoint operators or the Representation Theorem for semi-bounded sesquilinear forms are formulated but not proved. Where necessary we refer to the detailed exposition of the subject matter in the standard texts. The reader will find the basic principles set out in detail inter alia in the books [1], [26] and [56].

OPERATORS IN HILBERT SPACE

Let H be a separable, complex Hilbert space with an inner product and a norm denoted by

$$(u,v),\ \| u \|,\ u,v \in H\ .$$

In the following examination of spectral properties of differential operators the concrete Hilbert space $H = L_2(0, \infty)$, which is the set of (equivalence classes of) complex-valued measurable functions $u(x)$ with $\int_0^\infty |u(x)|^2 \,dx < \infty$, is principally considered; inner product and norm are then defined by

$$(u,v) = \int_0^\infty u(x)\,\bar{v}(x)\,dx,\ \| u \| = \left(\int_0^\infty |u(x)|^2 dx\right)^{\frac{1}{2}},\ u,v \in L_2(0,\infty)\ .$$

Occasionally the basic interval is the whole axis $(-\infty, \infty)$ or a finite interval.

We consider linear operators A whose domain $D(A)$ and range $R(A)$ are contained in H. The identity operator, with $D(E) = R(E) = H$, is denoted by E. A linear operator A with domain $D(A)$ dense in H is said to be symmetric if

$$(Au,v) = (u,Av), \; u,v \in D(A) \; ,$$

holds. If A is symmetric, then the adjoint operator A^* is an extension of A, and we can write $A \subseteq A^*$. A is said to be self-adjoint if A is equal to A^*. A symmetric operator is said to be essentially self-adjoint if the closure $\bar{A}$ of A is self-adjoint. Then the equation $\bar{A} = A^*$ holds and $D(\bar{A})$ arises from $D(A)$ by completion in the norm

$$(\| Au \|^2 + \| u \|^2)^{\frac{1}{2}} \; .$$

Regarding the stability of the self-adjointness or essential self-adjointness of an operator under perturbation by another operator, the following criterion of Rellich and Kato is fundamental.

Theorem 1.1

Let A be self-adjoint (essentially self-adjoint) and let B be symmetric with $D(B) \supseteq D(A)$. If the estimate

$$\| Bu \| \leqslant a \| u \| + b \| Au \|, \; a \geqslant 0, \; 0 \leqslant b < 1, \; u \in D(A) \; ,$$

holds, then $A + B$, where $D(A+B) = D(A)$, is self-adjoint (essentially self-adjoint).

See [26] for the proof.

A densely defined symmetric linear operator A is said to be semibounded from below if there exists a real number c such that

$$(Au,u) \geqslant c \| u \|^2, \; u \in D(A) \; .$$

In the case $c = 0$, A is said to be positive. If γ is chosen so that $c + \gamma > 0$, then the operator $A + \gamma E$, where $c_1 = c + \gamma$ and $D(A + \gamma E) = D(A)$, is positive definite, and this is expressed by the inequality

$$((A + \gamma E) u,u) \geqslant c_1 \| u \|^2, \quad c_1 > 0, \quad u \in D(A) \; .$$

Then by means of

$$[u,v]_A = ((A + \gamma E) u,v), \quad \| u \|_A = [u,u]_A^{\frac{1}{2}}, \quad u,v \in D(A) \; ,$$

an inner product and a norm are defined; $\| \cdot \|_A$ is the so-called energy norm of u with respect to A. If the linear set $D(A)$ is completed in the norm $\| \cdot \|_A$, then there arises a Hilbert space H_A, the so-called energy space H_A of A, which can be embedded in H; $H_A \subseteq H$. The notation $\| \cdot \|_A$ and H_A is justified because different γ with $c + \gamma > 0$ yield equivalent norms

$$((A + \gamma E)\, u,u)^{\frac{1}{2}} \;.$$

Each semibounded operator A with

$$(Au,u) \geqslant c\| u \|^2, \quad u \in D(A) \;,$$

possesses a special self-adjoint extension $\hat{A}$, called the Friedrichs extension of A, for which

$$(\hat{A}u,u) \geqslant c\| u \|^2, \quad u \in D(\hat{A}) \;,$$

also holds. The domain of $\hat{A}$ is given by

$$D(\hat{A}) = H_A \cap D(A^*) \;.$$

SPECTRAL THEOREM AND SPECTRUM

Each self-adjoint operator A is associated with a spectral family $\{E_\lambda\}_{-\infty < \lambda < \infty}$ of projections E_λ with the properties

$$\lim_{\lambda \to -\infty} E_\lambda u = 0 \text{ (zero vector of } H), \quad \lim_{\lambda \to \infty} E_\lambda u = u, \quad u \in H \;,$$

$$\lim_{\lambda \uparrow \mu} E_\lambda u = E_\mu u, \quad u \in H, \quad u \in (-\infty,\infty) \;,$$

$$E_\lambda E_\mu = E_\mu E_\lambda = E_{\min(\lambda,\mu)}, \quad \lambda,\mu \in (-\infty,\infty) \;.$$

By means of this spectral family the self-adjoint operator A can be represented as

$$Au = \int_{-\infty}^{\infty} \lambda \mathrm{d}E_\lambda u \;, \; D(A) = \{u \mid u \in H, \; \int_{-\infty}^{\infty} \lambda^2 \mathrm{d}\,(E_\lambda u,u) < \infty\} \;,$$

and $$\| Au \|^2 = \int_{-\infty}^{\infty} \lambda^2 \mathrm{d}\,(E_\lambda u,u) \;,$$

$$(Au,v) = \int_{-\infty}^{\infty} \lambda \, d\,(E_\lambda u,v), \quad u \in D(A), \quad v \in H \ .$$

(See [1] and [56]).

The spectrum $\sigma(A)$ of A is a set on the real axis which is determined by A and can be described by means of $\{E_\lambda\}_{-\infty < \lambda < \infty}$ as follows. A real number λ is an element of $\sigma(A)$ if and only if

$$E_{\lambda+\epsilon} - E_{\lambda-\epsilon} \neq 0 \quad \text{(zero operator)}$$

for every $\epsilon > 0$. If there exists $\epsilon > 0$ with

$$E_{\lambda+\epsilon} - E_{\lambda-\epsilon} = 0 \ ,$$

then the real number λ belongs to the resolvent set $\rho(A)$, being the complement of $\sigma(A)$ in the complex plane. Each complex number λ with Im $\lambda \neq 0$ belongs to $\rho(A)$. A real or complex λ belongs to $\rho(A)$ if and only if the operator $(A - \lambda E)^{-1}$ is bounded and defined on the whole of H. $\sigma(A)$ is a closed set of points on the real λ-axis. A point λ is said to be an eigenvalue of A if the equation $(A - \lambda E)u = 0$ possesses a nontrivial solution $u \in D(A)$. This is the case if and only if λ is real and

$$E_{\lambda+0} - E_\lambda \neq 0, \quad \text{where } E_{\lambda+0} = \lim_{\mu \downarrow \lambda} E_\mu \ .$$

Therefore eigenvalues belong to the spectrum; their totality forms the point spectrum $\sigma_p(A)$ of A. The dimension of the null space of $A - \lambda E$ is called the multiplicity of the eigenvalue λ. The union of the set of cluster points of the spectrum with the set of eigenvalues of infinite multiplicity forms the essential spectrum $\sigma_e(A)$. If no eigenvalues of infinite multiplicity are apparent – as is the case with the ordinary differential operators that concern us – then λ belongs to $\sigma_e(A)$ if and only if

$$E_{\lambda+\epsilon} - E_{\lambda+0} + E_\lambda - E_{\lambda-\epsilon} \neq 0$$

for every $\epsilon > 0$. The spectrum is said to be continuous on an open, half-open, or closed interval of the λ-axis if the interval is a subset of $\sigma(A)$ and contains no eigenvalue of A. Then on the interval in question $(E_\lambda u,u)$ is a continuous function of λ for each $u \in H$. The spectrum $\sigma(A)$ is called continuous if A possesses no eigenvalue. The spectrum is said to be discrete if it consists of isolated eigenvalues of finite multiplicity. This is the case if and only if $\sigma_e(A)$ is empty.

The points of $\sigma_e(A)$ can be characterized by Weyl sequences. A bounded and not pre-compact sequence

$$\{u_j\}_{j=1,2,\ldots}, \quad u_j \in D(A) \ ,$$

is called a Weyl sequence for the point λ with respect to A if

$$\lim_{j\to\infty} (Au_j - \lambda u_j) = 0 \ .$$

Then λ belongs to $\sigma_e(A)$ if and only if there exists a Weyl sequence for λ. Regarding the discreteness of the spectrum of a semibounded self-adjoint operator we have the following criterion due to Rellich.

Theorem 1.2
The spectrum of a self-adjoint operator which is semibounded below is discrete if and only if every bounded set in the energy space H_A is pre-compact in H.

See [56] for the proof.

SESQUILINEAR FORMS

A complex-valued function $a[u,v]$ defined in the product space $H \times H$ of a Hilbert space H, is said to be a sesquilinear form (in what follows it is also more briefly called a 'form'), if the elements u and v both vary in a linear subspace $D(a)$ of H and the following relationships hold:

$$a[\alpha_1 u_1 + \alpha_2 u_2, v] = \alpha_1 a[u_1, v] + \alpha_2 a[u_2, v] \ ,$$

$$a[u, \beta_1 v_1 + \beta_2 v_2] = \overline{\beta}_1 a[u, v_1] + \overline{\beta}_2 a[u, v_2] \ ,$$

$$\alpha_1, \alpha_2, \beta_1, \beta_2 \quad \text{complex} \ .$$

The form $a[u,v]$ is said to be symmetric if

$$a[u,v] = \overline{a[v,u]}, \quad u,v \in D(a) \ .$$

In this case the quadratic form $a[u,u]$, $u \in D(a)$, of the sesquilinear form $a[u,v]$ is real-valued.

A symmetric form $a[u,v]$ is said to be semibounded from below, if there exists a real number c such that

$$a[u,u] \geqslant c\| u \|^2, \quad u \in D(a) \tag{1.1}$$

If $a[u,v]$ is symmetric and semibounded below, then a sequence

$$\{u_j\}, \quad j=1,2,\ldots, \quad u_j \in D(a) ,$$

is said to be a-convergent to $u \in H$ if

$$\| u_j - u \| \to 0 \text{ as } j \to \infty \text{ and } a[u_j - u_k, u_j - u_k] \to 0 \text{ as } j,k \to \infty .$$

This property will be abbreviated as $u_j \underset{a}{\to} u (j \to \infty)$. If $u_j \underset{a}{\to} u (j \to \infty)$ implies that

$$u \in D(a) \text{ and } a[u_j - u, \; u_j - u] \to 0 \text{ as } j \to \infty ,$$

then the form $a[u,v]$ is said to be closed. A form $a[u,v]$, semibounded from below, is said to be closeable if there exists a closed form which is an extension of $a[u,v]$. Then there exists a smallest extension $\bar{a}[u,v]$ of $a[u,v]$ such that every closed extension of $a[u,v]$ is also an extension of $\bar{a}[u,v]$. The form $\bar{a}[u,v]$, with domain $D(\bar{a})$, is called the closure of $a[u,v]$. By means of a form $a[u,v]$ which satisfies (1.1), an inner product

$$[u,v]_a = a[u,v] + (1-c)(u,v), \quad u,v \in D(a) ,$$

is defined on $D(a)$. The completion of $D(a)$ in the norm

$$\| \cdot \|_a = [\,\cdot\,,\cdot\,]_a^{\frac{1}{2}}$$

yields a Hilbert space H_a. Then $D(a) = H_a$ if and only if $a[u,v]$ is closed. Concerning the closeability of a form, the following theorem holds.

Theorem 1.3
Let the form $a[u,v]$ be semibounded from below. Then $a[u,v]$ is closeable if and only if $u_j \underset{a}{\to} 0 \; (j \to \infty)$ implies that $a[u_j,u_j] \to 0 \; (j \to \infty)$. In this case the Hilbert space H_a may be considered as a subspace of H.

See [26] for the proof.

Thus the form

$$a[u,v] = (Au,v), \quad D(a) = D(A) ,$$

defined by means of an operator A semibounded from below, is closeable, and $H_a = H_A$.

Sesquilinear forms can be added. If the forms $a_1, \ldots, a_r$ are semibounded from below, then so is the form

$$a = a_1 + \ldots + a_r, \quad D(a) = D(a_1) \cap \ldots \cap D(a_r) .$$

If all the forms a_i, $i = 1, \ldots, r$, are closed (closeable), then the form a is also closed (closeable) and the inclusion

$$\bar{a} \subseteqq \bar{a}_1 + \ldots + \bar{a}_r$$

is valid.

In the sequel the following theorem will be used repeatedly.

Theorem 1.4
Let a and b be symmetric forms with $D(a) \subseteqq D(b)$ and $a[u,u] \geqslant 0, u \in D(a)$. If there exist real numbers α and β, with $0 \leqslant \beta < 1$, such that the estimate

$$| b[u,u] | \leqslant \alpha \| u \|^2 + \beta a[u,u], \quad u \in D(a) ,$$

holds, then:

(i) $a + b$, $D(a + b) = D(a)$, is semibounded from below.

(ii) $a + b$ is closed (closeable) if and only if a is closed (closeable).

(iii) if a is closeable, then $D(\overline{a + b}) = D(\bar{a})$.

See [26] for the proof.

We have remarked above that a semibounded operator defines a closeable form. Conversely, we have:

Theorem 1.5
Let the form $a[u,v]$ be densely defined, closed, and semibounded from below. There exists a unique self-adjoint operator A, semibounded from below, with the following properties:

1) $a[u,v] = (Au,v), u \in D(A), v \in D(a), D(A) \subseteqq D(a)$.

2) $D(A)$ is dense in $D(a)$ in the metric $\| \cdot \|_a$.

3) If, for any given $u \in D(a)$, the relationship $a[u,v] = (w,v)$ holds for some $w \in H$ and all v belonging to a dense subset of $D(a)$ in the metric $\| \cdot \|_a$, then $u \in D(a)$ and $Au = w$.

4) If $a[u,u] \geqslant 0, u \in D(a)$, then the equations

$$D(a) = D(A^{\frac{1}{2}}) \text{ and } a[u,v] = (A^{\frac{1}{2}}u, A^{\frac{1}{2}}v), \quad u,v \in D(a) ,$$

are valid.

In particular, a bounded form $a[u,v]$, $D(a) = H$, for which

$$| a[u,u] | \leqslant C \| u \|^2, \quad u \in D(a) = H ,$$

clearly defines a bounded self-adjoint operator A.

See [26] for the proof.

If A_0 is an operator which is semibounded from below and A is the self-adjoint operator defined according to Theorem 1.5 by the closure of

$$a_0[u,v] = (A_0 u,v), \quad u,v \in D(a_0) = D(A_0) \ ,$$

then A corresponds to the Friedrichs extension of A_0.

COURANT'S VARIATIONAL PRINCIPLE

A device often used in connection with the localization of the spectra of semi-bounded operators is Courant's Variational Principle. It goes as follows.

Theorem 1.6

Let $a[u,v]$ and $b[u,v]$ be sesquilinear forms which are closed and semibounded from below, and for which

$$D(a) \subseteqq D(b) \text{ and } a[u,u] \geqslant b[u,u] , \quad u \in D(a) \ .$$

Let A and B be the self-adjoint operators resulting from the forms in accordance with Theorem 1.5. If

$$\sigma_e(B) \cap (-\infty, \tilde{\lambda}) = \emptyset$$

for some $\tilde{\lambda}$, $-\infty < \tilde{\lambda} \leqslant \infty$, then also,

$$\sigma_e(A) \cap (-\infty, \tilde{\lambda}) = \emptyset \ .$$

Also, if the eigenvalues $\lambda_n(A)$ and $\lambda_n(B)$ of A and B which are smaller then $\inf \sigma_e(B)$ are numbered in increasing order with regard to their multiplicity, then

$$\lambda_n(A) \geqslant \lambda_n(B), \quad n = 1, 2, \ldots .$$

Proof [55]

Without loss of generality we suppose that A and B are positive. If $\{E_\lambda^{(A)}\}_{-\infty < \lambda < \infty}$ and $\{E_\lambda^{(B)}\}_{-\infty < \lambda < \infty}$ are the spectral families of A and B, then the equations

$$a[u,u] = \| A^{\frac{1}{2}} u \|^2 = \int_0^\infty \lambda \, \mathrm{d}(E_\lambda^{(A)} u,u), \quad u \in D(a) = D(A^{\frac{1}{2}}) \ ,$$

and $$b[u,u] = \|B^{\frac{1}{2}}u\|^2 = \int_0^\infty \lambda \, d(E_\lambda^{(B)}u,u), \quad u \in D(b) = D(B^{\frac{1}{2}}) \ ,$$

hold. As an indirect proof let us suppose the existence of a value λ' such that

$$\lambda_n(B) < \lambda' < \lambda_{n+1}(B)$$

and $$\dim E_{\lambda'}^{(B)}H = n < \dim E_{\lambda'}^{(A)}H \ .$$

$E_{\lambda'}^{(B)}H$ is spanned by an orthonormal basis of n eigenvectors

$$\phi_i, (\phi_i,\phi_j) = \delta_{ij}, \quad i,j = 1,\ldots,n \ ,$$

of B. Let $\psi_1, \ldots, \psi_n$ be $n+1$ linearly independent elements of

$$E_{\lambda'}^{(A)}H \subseteq D(A) \ .$$

By means of a linear combination of the $\psi_j, j = 1, \ldots, n+1$, an element ψ_0, $\|\psi_0\| = 1$, can be constructed such that ψ_0 is orthogonal to the subspace $E_{\lambda'}^{(B)}H$. Writing

$$\psi_0 = \sum_{j=1}^{n+1} c_j \psi_j \ ,$$

we obtain the system of equations

$$(\psi_0, \phi_i) = \sum_{j=1}^{n+1} c_j (\psi_j, \phi_i) = 0, \quad i = 1, \ldots, n \ ,$$

for the c_j, which has a nontrivial solution. Because

$$D(A) \subseteq D(a) \subseteq D(b) \ ,$$

ψ_0 belongs to $D(b)$ and, since $\psi_0 \perp E_{\lambda'}^{(B)}H$, we obtain

$$b[\psi_0, \psi_0] = \int_0^\infty \lambda \, d(E_\lambda^{(B)}\psi_0,\psi_0) = \int_{\lambda_{n+1}(B)}^\infty \lambda \, d(E_\lambda^{(B)}\psi_0,\psi_0)$$

$$\geqslant \lambda_{n+1}(B) \|\psi_0\|^2 = \lambda_{n+1}(B) \qquad (1.2)$$

On the other hand, $\psi_0 \in E_{\lambda'}^{(A)} H$ implies that

$$a[\psi_0,\psi_0] = \int_0^\infty \lambda \, \mathrm{d}(E_\lambda^{(A)}\psi_0,\psi_0) = \int_0^{\lambda'} \lambda \, \mathrm{d}(E_\lambda^{(A)}\psi_0,\psi_0) \leqslant \lambda' \, \| \psi_0 \|^2 = \lambda' \, ,$$

which together with (1.2), gives the estimate

$$a[\psi_0,\psi_0] \leqslant \lambda' < \lambda_{n+1}(B) \leqslant b[\psi_0,\psi_0] \, .$$

This contradicts the hypothesis of the theorem, and the theorem is thereby proved.

DECOMPOSITION OF OPERATORS

A further important device in the spectral theory of differential operators is the decomposition of operators [15]. Let a Hilbert space H be represented as an orthogonal sum $H = H_1 \oplus H_2$ of two subspaces H_1 and H_2. Let A_1 and A_2 be operators in H_1 and H_2 respectively, that is,

$$D(A_i) \subseteq H_i, \quad R(A_i) \subseteq H_i, \quad i = 1, 2 \, .$$

If the A_i are both self-adjoint, then so is the orthogonal sum

$$A = A_1 \oplus A_2, \quad Au = A_1 u_1 + A_2 u_2, \quad u = u_1 + u_2 \, ,$$

and we have

$$u_i \in D(A_i), \quad i = 1, 2 \, ,$$

Theorem 1.7

If $A = A_1 \oplus A_2$ is an orthogonal sum of self-adjoint operators, then

$$\sigma_{\mathrm{p}}(A) = \sigma_{\mathrm{p}}(A_1) \cup \sigma_{\mathrm{p}}(A_2) \text{ and } \sigma_{\mathrm{e}}(A) = \sigma_{\mathrm{e}}(A_1) \cup \sigma_{\mathrm{e}}(A_2) \, . \qquad (1.3)$$

Proof

The first equation is obvious. To prove the second we use Weyl sequences. Let $\lambda \in \sigma_{\mathrm{e}}(A_i)$, $i = 1$ or 2, and let $\{u_{ij}\}_{j=1,2,\ldots}$ be a corresponding Weyl sequence. Then $\{u_{i,j}\}_{j=1,2,\ldots}$ is also a Weyl sequence for λ with respect to A. Conversely, if $\lambda \in \sigma_{\mathrm{e}}(A)$ and $\{u_j\}_{j=1,2,\ldots}$ is a corresponding Weyl sequence, then

$$u_j = u_{1,j} + u_{2,j}, \quad u_{1,j} \in D(A_1), \quad u_{2,j} \in D(A_2), \quad j = 1, 2, \ldots \, ,$$

and either $\{u_{1,j}\}_{j=1,2,\ldots}$ is a Weyl sequence for λ with respect to A_1 or $\{u_{2,j}\}_{j=1,2,\ldots}$ is a Weyl sequence for λ with respect to A_2. For, from

$$Au_j - \lambda u_j \to 0, \quad j \to \infty \ ,$$

we obtain $A_i u_{i,j} - \lambda u_{i,j} \to 0, \ j \to \infty, \ i = 1, 2$,

and, if both sequences

$$\{u_{i,j}\}_{j=1,2,\ldots}, \quad i = 1, 2 \ ,$$

were pre-compact, then $\{u_j\}_{j=1,2,\ldots}$ would also be pre-compact, which is not the case. This proves Theorem 1.7.

DEFICIENCY INDICES

In this section, for future reference we set out some further theorems from the theory of extensions of symmetric operators [1]. A complex number λ is said to be a regular-type point of the linear operator A if there exists a $k = k(\lambda) > 0$ such that

$$\| (A - \lambda E) u \| \geqslant k \| u \|, \quad u \in D(A) \ .$$

Let A be a symmetric operator and let the complex number λ be written as $\lambda = \xi + i\eta$ with $\xi = \operatorname{Re} \lambda$ and $\eta = \operatorname{Im} \lambda$. Then a simple calculation shows that

$$\| (A - \lambda E) u \|^2 = \| (A - \xi E) u \|^2 + \eta^2 \| u \|^2 \geqslant \eta^2 \| u \|^2, \quad u \in D(A).$$

Therefore all λ with $\operatorname{Im} \lambda \neq 0$ are regular-type points of A. The set of all regular-type points of A is an open set in the complex plane and is called the domain of regularity of A. If A is symmetric, the domain of regularity is either connected – this is the case if there exists at least one real regular-type point – or it consists of the two half planes $\operatorname{Im} \lambda > 0$ and $\operatorname{Im} \lambda < 0$. In the interior of a connected component of the domain of regularity, the dimension of the orthogonal complement of $R(A - \lambda E)$ does not depend on λ. On the one hand this dimension is called the deficiency index $\operatorname{def} R(A - \lambda E)$ of the linear set $R(A - \lambda E)$ and on the other hand it is also called deficiency index of the operator A relative to the component in question. Accordingly a symmetric operator A possesses two deficiency indices namely

$$\operatorname{def} R(A - \lambda E) = \operatorname{def} R(A + iE) = m, \ \operatorname{Im} \lambda < 0 \ ,$$

and

$$\operatorname{def} R(A - \lambda E) = \operatorname{def} R(A - iE) = n, \ \operatorname{Im} \lambda > 0 \ ,$$

where m may be equal to n.

If A^* is the adjoint operator of the symmetric operator A, then each λ with $\operatorname{Im} \lambda > 0$ is an eigenvalue of multiplicity m of A^* and each λ with $\operatorname{Im} \lambda < 0$ is an eigenvalue of multiplicity n of A^*. Only if $m = n$ are there self-adjoint extensions of A.

For ordinary differential operators, which are considered later, the deficiency indices m and n are finite and equal. For such operators the following theorems are valid.

Theorem 1.8
All self-adjoint extensions of a symmetric operator A with finite and equal deficiency indices have the same essential spectrum.

See [1] for the proof.

Theorem 1.9
Let λ be a real regular-type point of the symmetric operator A with equal deficiency indices (m,m), $m < \infty$. Then there exists a self-adjoint extension $\tilde{A}$ of A for which λ is an eigenvalue of multiplicity m.

See [1] for the proof.

As regards self-adjoint extensions of operators which are semibounded from below, the following applies.

Theorem 1.10
If the spectrum of a self-adjoint extension of an operator with deficiency indices (m,m), $m < \infty$, is bounded below and discrete in $(-\infty, \lambda_0)$, then the same is true of the spectrum of any other self-adjoint extension.

See [1] for the proof.

Chapter 2

The Essential Spectrum

We now consider differential operators which are generated by the differential expression

$$a[\cdot] = \sum_{k=0}^{n} (-1)^k \frac{\mathrm{d}^k}{\mathrm{d}x^k} a_k(x) \frac{\mathrm{d}^k}{\mathrm{d}x^k}, \quad -\infty \leqslant x_0 < x < \infty, \quad n \geqslant 1 . \tag{1}$$

We make the following assumptions about the coefficients $a_k(x)$ in this chapter:

$$\begin{aligned} &\text{i) } a_k(x) \text{ is real-valued, } k = 0, \ldots, n, \text{ and } a_n(x) > 0, x_0 < x < \infty, \\ &\text{ii) } a_k(x) \in W^k_{2,\mathrm{loc}}(x_0, \infty), \; k = 0, \ldots, n. \end{aligned} \tag{2}$$

The second condition means that $a_k(x)$ belongs to the Sobolev space $W_2^k(x_1, x_2)$ for all $x_1, x_2, x_0 < x_1 < x_2 < \infty$. The Sobolev spaces of functions of one variable and the corresponding Embedding Theorem are discussed in the appendix. According to this theorem, the assumption

$$a_k(x) \in W_2^k(x_1, x_2)$$

implies that

$$a_k(x) \in C^{k-1}[x_1, x_2] .^\dagger$$

Furthermore the generalized first derivative of $a_k^{(k-1)}(x)$ exists and belongs to $L_2(x_1, x_2)$. This fact can also be interpreted as meaning that $a_k^{(k-1)}(x)$ is absolutely continuous on $[x_1, x_2]$ and the derivative of $a_k^{(k-1)}(x)$, which exists almost everywhere, belongs to $L_2(x_1, x_2)$. By means of

$$A_0 u = a[u], \quad u \in D(A_0) = C_0^\infty(x_0, \infty) ,$$

a symmetric operator A_0 is defined in $L_2(x_0, \infty)$ as can be seen easily by integration by parts.

†By $a_k(x) \in C^{k-1}[x_1, x_2]$ is meant that $a_k(x)$ is equivalent to a function which is continuously differentiable $k-1$ times on $[x_1, x_2]$.

THE ADJOINT OPERATOR

To ensure the existence of self-adjoint extensions of A_0, the essential spectrum of which concerns us in the present chapter, we prove that A_0 has equal deficiency indices. To do this we first determine the adjoint operator A_0^* of A_0.

Theorem 2.1

The adjoint operator of A_0 is defined by

$$A_0^* u = a[u] \; ,$$
$$D(A_0^*) = \{u \,|\, u \in W_{2,\mathrm{loc}}^{2n}(x_0, \infty) \cap L_2(x_0, \infty), a[u] \in L_2(x_0, \infty)\} \; . \tag{3}$$

Proof

Since the theorem states a well-known fact we will only sketch the proof. We refer to [46] where, by means of the quasi-derivatives

$$\begin{aligned} u^{[k]} &= u^{(k)}, \quad k = 0, \ldots, n-1, \\ u^{[n]} &= a_n(x)\, u^{(n)}, \\ u^{[n+k]} &= a_{n-k}(x)\, u^{(n-k)} - \frac{\mathrm{d}}{\mathrm{d}x} u^{[n+k-1]}, \quad k = 1, \ldots, n-1, \\ u^{[2n]} &= a_0(x)\, u - \frac{\mathrm{d}}{\mathrm{d}x} u^{[2n-1]} = a[u], \end{aligned} \tag{4}$$

an operator L is defined as follows. $D(L)$ is the set of all functions u in $L_2(x_0, \infty)$ whose quasi-derivatives up to order $2n-1$ are absolutely continuous and the quasi-derivative $u^{[2n]}$ belongs to $L_2(x_0, \infty)$. Then by means of

$$Lu = a[u], \; D(L) \; ,$$

a linear operator is defined in $L_2(x_0, \infty)$. L is the adjoint operator of

$$L_0' u = a[u] \; ,$$

$$D(L_0') = \{u \,|\, u \in D(L), \; u(x) = 0 \text{ when } x_0 < x < x_u \text{ or } x > X_u\} \; ,$$

as is shown in [46]. From (2) and (4) it follows that $D(L)$ can also be described thus: $D(L)$ is the set of all functions u in $L_2(x_0, \infty)$ whose derivatives up to order $2n-1$ are absolutely continuous and for which $a[u]$ belongs to $L_2(x_0, \infty)$. Hence, by the Embedding Theorem, Theorem A.2, it follows that $D(L)$ coincides with

$$D(L) = \{u \,|\, u \in W_{2,\mathrm{loc}}^{2n}(x_0, \infty) \cap L_2(x_0, \infty), \; a[u] \in L_2(x_0, \infty)\} \; .$$

Therefore we have to prove that $\overset{*}{A_0} = L$, where we know that $L = L_0'^*$. Obviously $D(A_0) \subseteqq D(L_0')$. Consequently we have only to prove that $D(L_0') \subseteqq D(\bar{A}_0)$, where $\bar{A}_0$ is the closure of A_0. $D(\bar{A}_0)$ originates from $D(A_0)$ through closure of $D(A_0)$ in the norm

$$(\| A_0 u \|^2 + \| u \|^2)^{\frac{1}{2}} \ . \tag{5}$$

Any chosen function of $D(L_0')$ possesses a support which is contained in a finite interval (x_1, x_2), $x_0 < x_1 < x_2 < \infty$. The function, when restricted to (x_1, x_2) belongs to $W_2^{2n}(x_1, x_2)$ and can accordingly be approximated by functions in $C_0^\infty(x_1, x_2)$ in the W_2^{2n}-norm. For functions whose support lies in (x_1, x_2) however, the norm (5) is equivalent to the W_2^{2n} norm, as can be seen easily from the estimate (A.9) in Theorem A.2†. Thus $D(L_0') \subseteqq D(\bar{A}_0)$, and Theorem 2.1 is proved.

In Chapter 1 it was noted that the deficiency indices of A_0 are the multiplicities of the eigenvalues in the equations

$$\overset{*}{A_0} u = iu, \quad \overset{*}{A_0} u = (-i)\, u \ .$$

By Theorem 2.1 these equations are differential equations

$$a[u] = iu, \quad a[u] = (-i)\, u \ , \tag{6}$$

the solutions of which are to be sought in the space

$$W_{2,\mathrm{loc}}^{2n}(x_0, \infty) \cap L_2(x_0, \infty) \ .$$

If u is a solution of one equation, then the complex conjugate $\bar{u}$ is a solution of the other, so that the deficiency indices of A_0 coincide. In [46] it is shown that the equations (6) can be treated like classical differential equations. Thus we obtain the inequality $0 \leqslant m \leqslant 2n$ for the number m of linearly independent $L_2(x_0, \infty)$ solutions of either equation (6). Thus we have established that A_0 has equal deficiency indices (m, m). Therefore by Theorem 1.8 we obtain

Theorem 2.2
All self-adjoint extensions of the operator

$$A_0 u = a[u], \quad D(A_0) = C_0^\infty(x_0, \infty), \quad -\infty \leqslant x_0 < \infty \ ,$$

have the same essential spectrum.

†We apply the estimate

$$\| u^{(k)} \|_{L_2(x_1, x_2)} \leqslant \epsilon \, \| u^{(l)} \|_{L_2(x_1, x_2)} + C_\epsilon \, \| u \|_{L_2(x_1, x_2)}, \ \epsilon > 0 \ ,$$

$$-\infty \leqslant x_1 < x_2 \leqslant \infty, \ 0 \leqslant k < l, \ u \in W_2^l(x_1, x_2), \ C_\epsilon = C(\epsilon, x_1, x_2, l) \ ,$$

which follows from (A.9), where ϵ (> 0) is arbitrary.

Since for such a self-adjoint operator there is no eigenvalue of infinite multiplicity, its essential spectrum coincides with the set of accumulation points of its spectrum.

DIFFERENTIAL OPERATORS WITH CONSTANT COEFFICIENTS SELF-ADJOINT AND ESSENTIALLY SELF-ADJOINT OPERATORS

In what follows we will first consider operators with constant coefficients and determine their essential spectra. Then, by perturbations of these operators, conclusions on the spectrum of more general operators can be obtained. The results achieved are generalizations of theorems by inter alia Birman [3], Glazman [15], Balslev [2], and Schechter [50] in the case of ordinary differential operators. As a preliminary, we prove the following theorem concerning the simplest operator of order $2n$.

Theorem 2.3

1) The Friedrichs extension of the operator

$$A_0 u = (-1)^n a_n u^{(2n)}, \quad a_n = \text{const.} > 0, \quad D(A_0) = C_0^\infty(0, \infty) ,$$

is the operator

$$\hat{A} u = (-1)^n a_n u^{(2n)} ,$$

$$D(\hat{A}) = \{u \,|\, u \in W_2^{2n}(0, \infty), \; u(0) = \ldots = u^{(n-1)}(0) = 0\} .$$

2) The restriction A_θ of the operator $\hat{A}$ to the set of the functions

$$D(A_\theta) = \{u \,|\, u \in C^\infty[0, \infty), \; u(x) = 0, \; x > x_u > 0, \; u(0) = \ldots = u^{(n-1)}(0) = 0\}$$

is essentially self-adjoint.

3) The operator

$$A_0 u = (-1)^n a_n u^{(2n)}, \quad a_n = \text{const.} > 0, \quad D(A_0) = C_0^\infty(-\infty, \infty) ,$$

is essentially self-adjoint. Its closure is the operator

$$\hat{A} u = (-1)^n a_n u^{(2n)}, \quad D(\hat{A}) = W_2^{2n}(-\infty, \infty) .$$

Proof

Since the assertions of the theorem belong to the well-known basic facts of spectral theory the proofs will not be given in detail.

1) $D(\hat{A})$ is determined by

$$D(\hat{A}) = D(\overset{*}{A_0}) \cap H_{A_0} ,$$

where $D(\overset{*}{A_0}) = W_2^{2n}(0, \infty)$ by Theorem 2.1. By the Embedding Theorem A.2, the functions in $W_2^{2n}(0, \infty)$ are equivalent to functions in $C^{2n-1}[0, \infty]$. The energy space H_{A_0} arises from $C_0^\infty(0, \infty)$ by completion in the norm

$$(\|a_n u^{(n)}\|^2 + \|u\|^2)^{\frac{1}{2}} ,$$

which is equivalent to the norm

$$\|u\|_{W_2^n(0, \infty)} .$$

Since, by the Embedding Theorem, convergence in $W_2^n(0, \infty)$ implies convergence in $C^{n-1}[0, \infty)$, the boundary conditions

$$u(0) = u'(0) = \ldots = u^{(n-1)}(0) = 0$$

are valid for all u in H_{A_0}. Therefore the functions in $D(\hat{A})$ are functions u in $W_2^{2n}(0, \infty)$ for which $u^{(\nu)}(0) = 0, \nu = 0, \ldots, n-1$.

We now prove that every function u in $W_2^{2n}(0, \infty)$ with $u^{(\nu)}(0) = 0, \nu = 0, \ldots, n-1$, belongs to $D(\hat{A})$. Since obviously such a function is an element of $D(\overset{*}{A_0})$ we have to show that it belongs to H_{A_0} also. The extended function

$$\tilde{u}(x) = \begin{cases} u(x), & 0 \leqslant x < \infty , \\ u^{(n)}(0)\, x^n/n! & -\infty < x < 0 , \end{cases}$$

belongs to $C^n(-\infty, \infty)$. With the smoothing (or mollifying) construction in Theorem A.4, a function

$$\vartheta(x) \in C^\infty(-\infty, \infty)$$

arises from

$$\chi(x) = \begin{cases} 0, -\infty < x < -\frac{1}{2} , \\ 1, -\frac{1}{2} \leqslant x < \infty , \end{cases}$$

by smoothing with the radius $\frac{1}{2}$. Then, defining

$$\tilde{u}_\eta(x) = \vartheta\left(\frac{x}{\eta}\right) \tilde{u}(x), \quad -\infty < x < \infty, \ \eta > 0 ,$$

we obtain a function $\tilde{u}_\eta(x)$ which is zero on $(-\infty, -\eta)$ and equal to $u(x)$ on $[0, \infty)$. The function

$$U_\eta(x) = \tilde{u}_\eta(x - 2\eta)\, \vartheta(\eta^{-1} - x), \quad 0 \leqslant x < \infty,$$

then belongs to $C_0^n(0, \infty)$. A simple calculation yields

$$\| u(x) - U_\eta(x) \|_{W_2^n(0,\infty)} \to 0, \ \eta \to 0 \ . \tag{7}$$

By smoothing with the radius $\eta' < \eta$ a function

$$U_{\eta,\eta'}(x) \in C_0^\infty(0, \infty)$$

arises from $U_\eta(x)$, $-\infty < x < \infty$, for which

$$\| u - U_{\eta,\eta'} \|_{W_2^n(0,\infty)} < \epsilon$$

for any $\epsilon > 0$, if first η and then η' are chosen sufficiently small. This is easily seen using Theorem A.4. Thus it is proved that $u(x)$ belongs to H_{A_0}. Hence $\hat{A}$ is the Friedrichs extension of A_0.

2) To prove the second assertion we extend each $u \in D(\hat{A})$ to $[-1, \infty)$ in such a way that the extended function $u^*(x)$ belongs to $W_2^{2n}(-1, \infty)$. The extension can be made by the polynomial

$$P_u(x) = \frac{u^{(n)}(0)}{n!} x^n + \ldots + \frac{u^{(2n-1)}(0)}{(2n-1)!} x^{2n-1}, \quad -1 \leqslant x \leqslant 0 \ .$$

The function

$$u^*(u) = \begin{cases} u(x), & 0 < x < \infty \ , \\ P_u(x), & -1 \leqslant x \leqslant 0 \ , \end{cases}$$

belongs to $C^{2n-1}[-1, \infty)$. The fact that

$$u^* \in W_2^{2n}(-1, \infty)$$

follows from the relationship

$$\int_{-1}^{\infty} \phi^{(2n)}(x)\, u^*(x)\, dx = \int_{-1}^{\infty} \phi(x)\, v(x)\, dx, \quad \phi(x) \in C_0^\infty(-1, \infty) \ ,$$

where
$$v(x) = \begin{cases} u^{(2n)}(x), & 0 < x < \infty , \\ 0, & -1 \leqslant x \leqslant 0 , \end{cases}$$

and is an element of $L_2(-1, \infty)$. By smoothing $u^*(x)$ with radius η, $0 < \eta < \frac{1}{2}$, we obtain a function

$$\overset{*}{u}_\eta(x) \in C^\infty(-\tfrac{1}{2}, \infty) .$$

By Theorem A.4, for any given $\epsilon > 0$, there exists a value

$$\eta_0(\epsilon), \quad 0 < \eta_0(\epsilon) < \tfrac{1}{2} ,$$

such that $| \overset{*}{u}{}_\eta^{(\nu)}(0) | < \epsilon, \quad \nu = 0, 1, \ldots, n-1, \; \eta < \eta_0(\epsilon)$.

The function

$$U_\eta(x) = \left(\overset{*}{u}_\eta(x) - \vartheta(-x) \sum_{\nu=0}^{n-1} \frac{\overset{*}{u}{}_\eta^{(\nu)}(0)}{\nu!} x^\nu \right) \vartheta(\eta^{-1} - x), \quad 0 \leqslant x < \infty ,$$

has the properties

$$U_\eta(x) \in C_0^\infty [0, \infty) \text{ and } U_\eta^{(\nu)}(0) = 0, \quad \nu = 0, \ldots, n-1 .$$

By Theorem A.4 we now obtain

$$\| u - U_\eta \|_{W_2^{2n}(0, \infty)} \to 0, \quad \eta \to 0 .$$

Therefore the completion of $D(A_\theta)$ in the norm

$$\| \cdot \|_{W_2^{2n}(0, \infty)}$$

contains $D(\hat{A})$. A_θ is closeable as a symmetric operator and the domain of the closure $\overline{A}_\theta$ of A_θ arises from $D(A_\theta)$ by completion in the norm

$$(\| A_\theta u \|^2 + \| u \|^2)^{\frac{1}{2}} = (\| a_n u^{(2n)} \|^2 + \| u \|^2)^{\frac{1}{2}} .$$

Since this norm is equivalent to

$$\| \cdot \|_{W_2^{2n}(0, \infty)} ,$$

A_θ is an extension of $\hat{A}$. As $\hat{A}$ is self-adjoint and $\overline{A}_\theta$ is symmetric we have $\overline{A}_\theta = \hat{A}$. Thus Assertion 2) is proved.

3) Obviously the closure of A_0 is $\hat{A}$. Since

$$D(A_0^*) = W_2^{2n}(-\infty, \infty),$$

the operator $\hat{A}$ is self-adjoint, and Theorem 2.3 is now proved.

By using Theorem 1.1, the respective self-adjointness and essential self-adjointness of operators arising from the operator of Theorem 2.3 by certain perturbations can be proved.

Theorem 2.4

1) If the conditions

$$\text{i)} \quad \inf_{0<x<\infty} a_n(x) > 0, \quad \sup_{0<x<\infty} a_n(x) < \infty ,$$

$$\text{ii)} \quad \sup_{0<x<\infty} \int_x^{x+1} ([a_k^{(k)}(t)]^2 + [a_k(t)]^2)\, \mathrm{d}t < \infty, \quad 0 \leqslant k \leqslant n , \tag{8}$$

are satisfied, then the operator

$$Au = a[u],\ D(A) = \{u \mid u \in W_2^{2n}(0, \infty),\ u(0) = \ldots = u^{(n-1)}(0) = 0\}$$

is self-adjoint. The restricted operator A_θ with the domain

$$D(A_\theta) = \{u \mid u \in C_0^\infty [0, \infty),\ u(0) = \ldots = u^{(n-1)}(0) = 0\}$$

is essentially self-adjoint.

2) If the conditions

$$\text{i)} \quad \inf_{-\infty<x<\infty} a_n(x) > 0, \quad \sup_{-\infty<x<\infty} a_n(x) < \infty ,$$

$$\text{ii)} \quad \sup_{-\infty<x<\infty} \int_x^{x+1} ([a_k^{(k)}(t)]^2 + [a_k(t)]^2)\, \mathrm{d}t < \infty, \quad 0 \leqslant k \leqslant n ,$$

are satisfied, then the operator

$$A_0 u = a[u],\ D(A_0) = C_0^\infty(-\infty, \infty) ,$$

is essentially self-adjoint, and the operator

$$Au = a[u],\ D(A) = W_2^{2n}(-\infty, \infty) ,$$

is self-adjoint.

Proof

1) We define α_1, α_2, and α by

$$\inf_{0<x<\infty} a_n(x) = \alpha_1, \quad \sup_{0<x<\infty} a_n(x) = \alpha_2, \quad \alpha = \tfrac{1}{2}(\alpha_1 + \alpha_2) \ ,$$

and define the operators

$$Tu = (-1)^n \alpha u^{(2n)}, \quad D(T) = D(A) \ ,$$

and

$$Bu = (-1)^n [(a_n(x) - \alpha) u^{(n)}]^{(n)} + \sum_{k=0}^{n-1} (-1)^k (a_k(x) u^{(k)})^{(k)} \ ,$$

$$D(B) = D(A) \ .$$

The operator T is self-adjoint by Theorem 2.3. We prove that B is symmetric and that an estimate of the form

$$\| Bu \| \leqslant a \| u \| + b \| Tu \|, \ 0 \leqslant a, \ 0 \leqslant b < 1, \ u \in D(T) \ , \tag{9}$$

holds. We begin with the proof of (9). There is a constant $C > 0$ such that

$$\| Bu \| \leqslant \| (a_n(x) - \alpha) u^{(2n)} \| + C \sum_{\substack{0 \leqslant \nu \leqslant k \leqslant n \\ 2k - \nu < 2n}} \| a_k^{(\nu)}(x) u^{(2k-\nu)} \| \tag{10}$$

for $u \in D(B)$. As an estimate for the first term on the right-hand side of (10) we have

$$\| (a_n(x) - \alpha) u^{(2n)} \| \leqslant \tfrac{1}{2}(\alpha_2 - \alpha_1) \| u^{(2n)} \| = (\alpha_2 - \alpha_1)(\alpha_2 + \alpha_1)^{-1} \| Tu \| \ . \tag{11}$$

Concerning the other terms the application of (8) and (A.9) gives

$$\| a_k^{(\nu)} u^{(2k-\nu)} \|^2 = \int_0^\infty [a_k^{(\nu)}(x)]^2 \, | u^{2k-\nu}(x) |^2 \, \mathrm{d}x$$

$$= \sum_{m=0}^{\infty} \int_m^{m+1} [a_k^{(\nu)}(x)]^2 \, | u^{(2k-\nu)}(x) |^2 \, \mathrm{d}x$$

$$\leqslant \sum_{m=0}^{\infty} \left(\max_{m \leqslant x \leqslant m+1} | u^{(2k-\nu)}(x) |^2 \right) \int_m^{m+1} | a_k^{(\nu)}(x) |^2 \, \mathrm{d}x$$

$$\leqslant C \sum_{m=0}^{\infty} \max_{m \leqslant x \leqslant m+1} | u^{(2k-\nu)}(x) |^2$$

$$\leqslant C \sum_{m=0}^{\infty} \left(\epsilon \int_m^{m+1} | u^{(2n)} |^2 \, dx + C_\epsilon \int_m^{m+1} |u|^2 \, dx \right) \leqslant C \epsilon \| u^{(2n)} \|^2 + C C_\epsilon \| u \|^2 .$$

Therefore for any given $\epsilon' > 0$, there exists a $C_{\epsilon'}$, such that

$$\| a_k^{(\nu)} u^{(2k-\nu)} \| \leqslant \epsilon' \| u^{(2n)} \| + C_{\epsilon'} \| u \| \leqslant \epsilon' \alpha^{-1} \| Tu \| + C_{\epsilon'} \| u \| , \quad (12)$$

$$k \leqslant 2k - \nu < 2n .$$

If ϵ' is chosen sufficiently small we obtain the inequality (9) with $b < 1$ from (10), (11), and (12).

We now show that B is symmetric. Let

$$\tilde{a}_n(x) = a_n(x) - \alpha, \; \tilde{a}_k(x) = a_k(x), \quad k = 0, \ldots, n-1 .$$

By integration by parts we obtain

$$(Bu, u) = \lim_{\substack{x_1 \to 0 \\ x_2 \to \infty}} \left(\sum_{k=0}^{n} \int_{x_1}^{x_2} \tilde{a}_k(x) \, | u^{(k)}(x) |^2 \, dx + \right.$$

$$\left. \sum_{k-1}^{n} \sum_{\substack{\lambda_k + \mu_k + \nu_k = 2k-1 \\ 0 \leqslant \lambda_k < k \\ 0 \leqslant \nu_k < k}} C_{\lambda_k \mu_k \nu_k} \left[a_k^{(\lambda_k)}(x_2) \, u^{(\mu_k)}(x_2) \, \bar{u}^{(\nu_k)}(x_2) - a_k^{(\lambda_k)}(x_1) \, u^{(\mu_k)}(x_1) \, \bar{u}^{(\nu_k)}(x_1) \right] \right)$$

with certain constants $C_{\lambda_k \mu_k \nu_k}$. To estimate a product

$$a_k^{(\lambda_k)}(x_2) \, u^{(\mu_k)}(x_2) \, \bar{u}^{(\nu_k)}(x_2)$$

we use (A.9) and so we obtain

$$| a_k^{(\lambda_k)}(x_2) \, u^{(\mu_k)}(x_2) \, \bar{u}^{(\nu_k)}(x_2) | \leqslant C \, \| a_k \|_{W_2^k(x_2, x_2+1)} \, \|u\|^2_{W_2^{2n}(x_2, x_2+1)} {}^{\dagger}$$

By condition (8) we have

$$\|a_k\|_{W_2^k(x_2, x_2+1)} \leqslant C, \quad 0 < x_2 < \infty ,$$

†Positive constants appearing in estimates are often denoted by the same capital letter C.

and, since $u \in W_2^{2n}(0, \infty)$, the factor

$$\| u \|^2_{W_2^{2n}(x_2, x_2+1)}$$

tends to zero when $x_2 \to \infty$. The products

$$a_k^{(\lambda_k)}(x_1)\, u^{(\mu_k)}(x_1)\, \bar{u}^{(\nu_k)}(x_1)$$

similarly tend to zero when $x_1 \to 0$ since, by supposition, we have

$$u^{(\nu)}(0) = 0, \; \nu = 0, \ldots, \; n-1 \;,$$

and the factors $a_k^{(\lambda_k)}(x_1)$ are bounded functions. Hence from (13) it follows that

$$(Bu, u) = \sum_{k=0}^{n} \int_0^\infty \tilde{a}_k(x)\, | u^{(k)} |^2 \, dx, \quad u \in D(B) \;.$$

Since (Bu, u) is real-valued, B is symmetric. Because in addition the estimate (9) holds, it follows from Theorem 1.1 that the operator $A = T + B$ is self-adjoint.

The second assertion in 1) follows similarly if we use the fact that the unperturbed operator

$$T_\theta u = (-1)^n \alpha\, u^{(2n)}, \; D(T_\theta) = D(A_\theta) \;,$$

is essentially self-adjoint by Theorem 2.3.

2) The proof is similar and is therefore omitted. Thus Theorem 2.4 is proved.

We now turn to the determination of the essential spectrum. The following theorem on the localization of the essential spectrum of the operator with constant coefficients is the basis for corresponding theorems for more general operators.

Theorem 2.5

All self-adjoint extensions of the symmetric operator

$$A_0 u = \sum_{k=0}^{n} (-1)^k a_k\, u^{(2k)}, \quad D(A_0) = C_0^\infty(0, \infty) \;,$$

$$a_k = \text{const.}, \quad k = 0, \ldots, n, \; a_n > 0 \;,$$

have the same essential spectrum which is the interval

$$[\Lambda, \infty), \quad \Lambda = \inf_{0<\xi<\infty} \sum_{k=0}^{n} a_k \, \xi^{2k} \; .$$

The essential spectrum of the self-adjoint operator

$$Au = \sum_{k=0}^{n} (-1)^k a_k \, u^{(2k)}, \; D(A) = W_2^{2n}(-\infty, \infty) \; ,$$

is also $[\Lambda, \infty)$.

Proof

Since by Theorem 2.2 all self-adjoint extensions of A_0 have the same essential spectrum, it is sufficient to consider the Friedrichs extension $\hat{A}$ of A_0. We first show that the interval $(-\infty, \Lambda)$ contains no point of the spectrum of $\hat{A}$. To do this we use the Fourier transform (see [56])

$$\phi(\xi) = (2\pi)^{-\frac{1}{2}} \int_{-\infty}^{\infty} \mathrm{e}^{-ix\xi} \, \phi(x) \, \mathrm{d}x, \quad \phi(x) \in C_0^{\infty}(-\infty, \infty) \; .$$

This transform is isometric in $L_2(-\infty, \infty)$ and has the property

$$(-i)^k \, \widehat{\phi^{(k)}} = \xi^k \hat{\phi} \; .$$

If a function $u \in D(A_0)$ is extended to the whole x-axis by means of $u = 0$ in $(-\infty, 0)$ and if the extended function is denoted by v, then using the Fourier transform we have

$$(A_0 u, u) = \sum_{k=0}^{n} (-1)^k \int_{-\infty}^{\infty} a_k \, v^{(2k)}(x) \, \bar{v}(x) \, \mathrm{d}x$$

$$= \sum_{k=0}^{n} \int_{-\infty}^{\infty} a_k \, \xi^{2k} \, |\hat{v}(\xi)|^2 \, \mathrm{d}\xi \geqslant \Lambda \int_{-\infty}^{\infty} |\hat{v}(\xi)|^2 \, \mathrm{d}\xi = \Lambda \, \|u\|^2 \; .$$

By the definition of the Friedrichs extension, it is also true that

$$(\hat{A}u, u) \geqslant \Lambda \, \|u\|^2, \quad u \in D(\hat{A})$$

and therefore the spectrum of A lies in the interval $[\Lambda, \infty)$.

We have still to prove that every point of $[\Lambda, \infty)$ belongs to $\sigma_e(A)$. For this purpose a Weyl sequence will be constructed for each $\lambda \in [\Lambda, \infty)$. There exists a ξ such that

$$\sum_{k=0}^{n} a_k \xi^{2k} = \lambda . \qquad (14)$$

Let the intervals

$$I_m = \{x \mid |x - x_m| < l_m\}, \quad l_m \to \infty, \quad m \to \infty$$

lie in $(0, \infty)$ and be mutually disjoint. By smoothing with the radius $\frac{1}{2}$ we obtain a function $\eta(x) \in C_0^\infty(-\infty, \infty)$ from

$$\chi(x) = \begin{cases} 0, & |x| > \frac{1}{2} \\ 1, & |x| \leqslant \frac{1}{2} , \end{cases}$$

the support of $\eta(x)$ being the interval $[-1, 1]$. Then the functions

$$u_m(x) = l_m^{-\frac{1}{2}} \eta[(x - x_m) l_m^{-1}] \exp[i\xi(x - x_m)], \quad m = 1, 2, \ldots ,$$

have the property

$$(u_m, u_{m'}) = 0, \quad m \neq m' .$$

By (14) they form a Weyl sequence for the point $\lambda \in [\Lambda, \infty)$ as can be seen from a simple calculation.

The essential spectrum of the self-adjoint operator

$$Au = \sum_{k=0}^{n} (-1)^k a_k u^{(2k)}, \; D(A) = W_2^{2n}(-\infty, \infty), \; a_k = \text{const.}, a_n > 0 ,$$

clearly coincides with the interval $[\Lambda, \infty)$ also. This proves Theorem 2.5.

INVARIANCE OF THE ESSENTIAL SPECTRUM UNDER PERTURBATIONS

To describe the essential spectrum of more general operators we now use theorems of Birman from the theory of perturbations of quadratic forms. Perturbations by operators which do not change the essential spectrum and the order of which is higher than the order of the unperturbed operator have been considered by Schechter [49], [50]. In what follows, these perturbations are considered also, and for this purpose we generalize a theorem by Birman [3], the proof in question (see [15]) needing only slight modification.

Theorem 2.6
Let A_0 and B_0 be positive definite operators with $D(A_0) = D(B_0)$ such that

$$\| u \|^2 \leqslant (B_0 u, u) \leqslant c(A_0 u, u), \quad u \in D(B_0) \ , \tag{15}$$

where c is a positive constant. If the Friedrichs extensions are denoted by $\hat{A}$ and $\hat{B}$ and the energy spaces by H_{A_0} and H_{B_0} respectively, then we assume that $D(\hat{B}) \subseteqq H_{A_0}$. As a condition on the operator $S_0 = A_0 - B_0$, let there exist a form

$$s[u, v], \ D(s) = H_{A_0} \ ,$$

with the properties

$$1) \ 0 < s[u, u], \quad u \neq 0 \ , \tag{16}$$

$$2) \ s[u, u] \leqslant C \| u \|^2_{A_0}, \quad C > 0, \quad u \in H_{A_0} \tag{17}$$

$$3) \ |(S_0 u, u)| \leqslant s[u, u], \quad u \in D(S_0) \ . \tag{18}$$

Also, let the boundedness of a set $M \subseteqq H$ imply the precompactness of the image $\hat{B}^{-1} M$ in the metric $s^{\frac{1}{2}}[u, u]$. Then

$$\sigma_e(\hat{A}) = \sigma_e(\hat{B}) \ .$$

Proof
By (17) and (18) the form $(S_0 u, v)$ is bounded in H_{A_0}. Therefore by Theorem 1.5 there exists a bounded self-adjoint operator T in H_{A_0} such that

$$(S_0 u, v) = (\hat{A}\hat{A}^{-1} S_0 u, v) = [\hat{A}^{-1} S_0 u, v]_{A_0} = [Tu, v]_{A_0} \ , \tag{19}$$

$$u \in D(B_0), \quad v \in H_{A_0} \ .$$

Now, by (19), we have

$$(\hat{A}u, v) - (\hat{B}u, v) = [Tu, v]_{A_0}$$

or

$$[u, v]_{A_0} - [u, v]_{B_0} = [Tu, v]_{A_0}, \quad u \in D(B_0), \quad v \in H_{A_0} \ . \tag{20}$$

Since $\|\cdot\|_{A_0}$-convergence implies $\|\cdot\|_{B_0}$-convergence and the operator T and the inner products $[\,.\,,.\,]$ are continuous, the range of u in (20) may be extended to H_{A_0}. Therefore

$$[u, v]_{A_0} - [u, v]_{B_0} = [Tu, v]_{A_0}, \quad u, v \in H_{A_0} \tag{21}$$

also holds. If for any $h \in H$ we define

$$\phi = \hat{A}^{-1}h \quad \text{and} \quad \psi = \hat{B}^{-1}h \ ,$$

then ψ belongs to H_{A_0} since $D(\hat{B}) \subseteqq H_{A_0}$. Now, from

$$[\psi, v]_{B_0} = (\hat{B}\psi, v) = (h, v) = (\hat{A}\phi, v) = [\phi, v]_{A_0}$$

and (21) with $u = \psi$, we obtain

$$[\psi, v]_{A_0} - [\phi, v]_{A_0} = [(\hat{B}^{-1} - \hat{A}^{-1})h, v]_{A_0} = [T\hat{B}^{-1}h, v]_{A_0} \ .$$

Hence $\qquad (\hat{B}^{-1} - \hat{A}^{-1})h = T\hat{B}^{-1}h, \quad h \in H \ .$

In what follows we prove that the operator $T\hat{B}^{-1}$ is compact in H. Then, by a theorem of Weyl, the compactness of

$$T\hat{B}^{-1} = \hat{B}^{-1} - \hat{A}^{-1}$$

implies that $\hat{A}^{-1}$ and $\hat{B}^{-1}$ have the same essential spectrum from which, finally, we obtain

$$\sigma_e(\hat{A}) = \sigma_e(\hat{B}) \ .$$

As a side issue we prove that the inequality

$$|(S_0u, u)| \leqslant s[u, u], \quad u \in D(S_0) \ ,$$

given in (18), implies that

$$|(S_0u, v)|^2 \leqslant s[u, u] \cdot s[v, v], \quad u \in D(S_0), \quad v \in H_{A_0} \ . \tag{22}$$

We have

$$(S_0u, w) + (S_0w, u) = \tfrac{1}{2}[(S_0(u + w), \mu + w) - (S_0(u - w), u - w)] \ ,$$
$$u, w \in D(S_0) \ ,$$

and hence

$$|(S_0u, w) + (S_0w, u)| \leqslant \tfrac{1}{2}(s[u+w, u+w] + s[u-w, u-w]) \qquad (23)$$
$$= s[u, u] + s[w, w] \; .$$

First we suppose that

$$s[u, u] = 1, \quad s[w, w] = 1 \quad \text{and} \quad w = e^{i\alpha} v \; ,$$

where α is defined by

$$(S_0u, v) = |(S_0u, v)| \, e^{i\alpha}, \quad (S_0u, v) \neq 0 \; .$$

Then (23) gives

$$|(S_0u, v)| \leqslant 1 \; . \qquad (24)$$

If we now have, more generally,

$$s[u, u] > 0, \quad s[w, w] > 0 \; ,$$

then the inequality

$$|(S_0u, v)|^2 \leqslant s[u, u] \cdot s[v, v] \qquad (25)$$

follows from (24). Here (25) holds for all $u, v \in D(S_0)$ because of (16), and, finally, v can also be an element of H_{A_0}.

If we put $v = \hat{A}^{-1}S_0u$ in (22) and use (17), then we obtain

$$(\hat{A}v, v)^2 \leqslant s[u, u] \cdot s[v, v] \leqslant C\, s[u, u] \, (\hat{A}v, v) \; .$$

Hence $\qquad (\hat{A}v, v) \leqslant C\, s[u, u], \quad v = \hat{A}^{-1}S_0u \; ,$

that is, $\qquad [v, v]_{A_0} \leqslant C\, s[u, u], \quad v = Tu, \quad u \in D(S_0) \; ,$

since $\qquad \hat{A}^{-1}S_0u = Tu, \quad u \in D(S_0) \; ,$

by (19). Since T and the form $s[u, v]$ are continuous in H_{A_0}, we also have

$$[v, v]_{A_0} \leqslant C\, s[u, u], \quad v = Tu, \quad u \in H_{A_0} \; . \qquad (26)$$

Now let M be a bounded set in H. Then by hypothesis the image $\hat{B}^{-1} M$ is s-pre-compact, and hence by (26) the set

$$\{v \mid v = Tu = T\hat{B}^{-1}h, \quad h \in M\}$$

is pre-compact in H_{A_0}. Therefore the set $T\hat{B}^{-1}M$ is also pre-compact in H, so that the compactness of

$$T\hat{B}^{-1} = \hat{B}^{-1} - \hat{A}^{-1}$$

is proved. This completes the proof of Theorem 2.6.

Note 2.7

Theorem 2.6 is also true if, in the hypothesis, two closed positive definite forms $a[u, v]$ and $b[u, v]$, for which

$$\| u \|^2 \leqslant b[u, u] \leqslant c\, a[u, u], \quad u \in D(a), \quad c > 0 ,$$

are used in place of the operators A_0 and B_0. Then $\hat{A}$ and $\hat{B}$ are the self-adjoint operators defined by the forms $a[u, v]$ and $b[u, v]$. In place of $(S_0 u, v)$ we use the form

$$s_0[u, v] = a[u, v] - b[u, v], \; D(s_0) = D(a) = H_{\hat{A}} \; .$$

Since, however, in this chapter the coefficients $a_k(x)$ are assumed to be so smooth that the self-adjoint operators in question are extensions of the symmetric operators defined on $C_0^\infty(0, \infty)$ and $C_0^\infty(-\infty, \infty)$ respectively, we continue to refer to the situation formulated in Theorem 2.6.

In later applications to concrete differential operators the next simple theorem provides a sharpening of the results obtained by means of Theorem 2.6.

Theorem 2.8

Let $\{A_{\nu,0}\}_{\nu=1,2,\ldots}$ be a sequence of positive definite operators and let $\{\hat{A}_\nu\}_{\nu=1,2,\ldots}$ be the sequence of the correponding Friedrichs extensions. Let

$$\sigma_e(\hat{A}_\nu) = \sigma_e(\hat{A}_1), \; \nu = 1, 2, \ldots , \tag{27}$$

and let B be a positive definite self-adjoint operator such that

$$\| B^{-1} - \hat{A}_\nu^{-1} \| \to 0, \quad \nu \to \infty . \tag{28}$$

Then $\quad \sigma_e(B) = \sigma_e(\hat{A}_1)$.

Proof
From (27) we have

$$\sigma_e(\hat{A}_\nu^{-1}) = \sigma_e(\hat{A}_1^{-1}), \quad \nu = 1, 2, \ldots .$$

We will show that

$$\sigma_e(B^{-1}) = \sigma_e(\hat{A}_1^{-1}) .$$

For any $\epsilon > 0$ there exists a value ν_0 such that

$$\| B^{-1} - \hat{A}_{\nu_0}^{-1} \| < \epsilon . \tag{29}$$

Then the set $\sigma_e(B^{-1})$ lies in the ϵ-neighbourhood of the set $\sigma_e(A_{\nu_0}^{-1})$, as can be seen as follows. Suppose on the contrary that there exists $\mu \in \sigma_e(B^{-1})$ with

$$\text{dist}\,(\mu, \sigma_e(\hat{A}_{\nu_0}^{-1}))\| \leqslant \epsilon . \tag{30}$$

and let $\{u_m\}_{m=1,2,\ldots}, \ (u_m, u_{m'}) = \delta_{m,m'}$,

be a Weyl sequence for μ in respect of B^{-1} with

$$u_m \in (E_{\mu+1/m}^{(B^{-1})} - E_{\mu-1/m}^{(B^{-1})}) H, \quad m = 1, 2, \ldots , \tag{31}$$

Where $\{E_\lambda^{(B^{-1})}\}_{-\infty<\lambda<\infty}$ denotes the spectral family of B^{-1}. If η is any number with $0 < \eta < \epsilon$, then by (30) there are at most a finite number of points of $\sigma(\hat{A}_{\nu_0}^{-1})$ in the interval $[\mu - \eta, \mu + \eta]$. These points are eigenvalues of $\hat{A}_0^{-1}$, and the corresponding eigenvectors span a finite dimensional subspace $H_{(\mu,\eta)}$. As the subspaces (31) are all of infinite dimension, the u_m can be chosen to be orthogonal to $H_{(\mu,\lambda)}$. Then we have

$$\| \hat{A}_{\nu_0}^{-1} u_m - \mu u_m \|^2 = \int_{-\infty}^{\mu-\eta} (\lambda - \mu)^2 \, d(E_\lambda^{(\hat{A}_{\nu_0}^{-1})} u_m, u_m)$$

$$+ \int_{\mu+\lambda}^{\infty} (\lambda - \mu)^2 \, d(E_\lambda^{(\hat{A}_{\nu_0}^{-1})} u_m, u_m) \geqslant \eta^2 .$$

Now, from

$$\eta \leqslant \| \hat{A}_{\nu_0}^{-1} u_m - \mu u_m \| \leqslant \| B^{-1} u_m - \mu u_m \| + \| (\hat{A}_{\nu_0}^{-1} - B^{-1}) u_m \|$$

$$\leqslant \| B^{-1} u_m - \mu u_m \| + \| B^{-1} - \hat{A}_{\nu_0}^{-1} \|$$

and by letting $m \to \infty$, we obtain

$$\eta \leqslant \| B^{-1} - \hat{A}_{\nu_0}^{-1} \| \ .$$

This contradicts (29) if we choose, for example,

$$\eta = \tfrac{1}{2}(\epsilon + \| B^{-1} - \hat{A}_{\nu_0}^{-1} \|) \ .$$

Thus there is no point $\mu \in \sigma_e(B^{-1})$ such that (30) holds, and therefore $\sigma_e(B^{-1})$ lies in the ϵ-neighbourhood of

$$\sigma_e(\hat{A}_{\nu_0}^{-1}) = \sigma_e(\hat{A}_1^{-1}) \ .$$

If we interchange the roles of B^{-1} and $\hat{A}_{\nu_0}^{-1}$, then it follows similarly that the set $\sigma_e(\hat{A}_1^{-1})$ also lies in the ϵ-neighbourhood of $\sigma_e(B^{-1})$. Since ϵ is chosen arbitrarily and the essential spectra are closed sets, we have

$$\sigma_e(B^{-1}) = \sigma_e(\hat{A}_1^{-1})$$

and consequently

$$\sigma_e(B) = \sigma_e(A_1)$$

as required.

LOCALIZATION OF THE ESSENTIAL SPECTRUM

We begin by applying Theorem 2.6 to concrete differential operators.

Theorem 2.9

Let the coefficients $a_k(x)$ of the operator

$$A_0 u = \sum_{k=0}^{n^*} (-1)^k \, (a_k(x)\, u^{(k)})^{(k)}, \quad D(A_0) = C_0^\infty(0, \infty), \quad n \leqslant n^* < 2n \ ,$$

satisfy the following conditions.

1) The $a_k(x)$ are real-valued and

$$a_k(x) \in W_2^k(0, X), \quad k = 0, \ldots, n^*, \quad \text{for every } X > 0 \ ,$$

and
$$\inf_{0<x<\infty} a_n(x) > 0 \ .$$

2) There are constants b_k, such that

$$\lim_{x\to\infty}\int_x^{x+1} |a_k(t) - b_k|\,\mathrm{d}t = 0,\quad k = 0,\ldots,n\ .$$

Further, in the case $n^* > n$, let

$$a_k(x) \geqslant 0,\quad \lim_{x\to\infty}\int_x^{x+1} a_k(t)\,\mathrm{d}t = 0,\quad k = n+1,\ldots,n^*\ ,$$

and $$a_{n^*}(x) > 0,\quad 0 < x < \infty\ .$$

Then the operator A_0 is semibounded from below, and the essential spectrum of every self-adjoint extension of A_0 is the interval $[\Lambda, \infty)$, where

$$\Lambda = \inf_{0<\xi<\infty}\ \sum_{k=0}^{n} b_k\,\xi^{2k}\ .$$

Proof

The proof will be developed in several stages. The first four stages are preparatory and deal with the construction of an operator ${}_*A_0$ with the same essential spectrum as A_0 and to which Theorem 2.6 can be applied. The basic interval of this new operator is $(-\infty, \infty)$.

1) First we simplify the functions $a_k(x)$ in a neighbourhood of $x = 0$ and thus define a new operator

$${}_+A_0,\ D({}_+A_0) = C_0^\infty(0,\infty)\ ,$$

the self-adjoint extensions of which possess the same essential spectrum as the self-adjoint extensions of A_0. By smoothing the function

$$\chi(x) = \begin{cases} 0, & -\infty < x \leqslant \tfrac{3}{2}\ , \\ 1, & \tfrac{3}{2} < x < \infty\ , \end{cases}$$

with radius $\tfrac{1}{2}$, we obtain the function

$$\phi(x) \begin{cases} = 0, & -\infty < x \leqslant 1\ , \\ = 1, & 2 \leqslant x < \infty,\ 0 < \phi(x) < 1,\ 1 < x < 2\ , \\ \in C_0^\infty(-\infty,\infty)\ , \end{cases}$$

by means of which the new coefficients are defined as follows:

$$
\begin{aligned}
{}_+a_{n^*}(x) &= 1 + \phi(x)[a_{n^*}(x) - 1] \ , \\
{}_+a_n(x) &= 1 + \phi(x)[a_n(x) - 1] \ , \\
{}_+a_k(x) &= \phi(x)a_k(x), \ \ k \neq n, n^* \ .
\end{aligned}
$$

Then we have

$$\inf_{0<x<\infty} \ {}_+a_n(x) = \alpha > 0 \tag{32}$$

and $\qquad {}_+a_{n^*}(x) > 0, \ \ 0 < x < \infty$.

The operator

$${}_+A_0u = \sum_{k=0}^{n^*} (-1)^k \ ({}_+a_k(x)u^{(k)})^{(k)}, \ \ D({}_+A_0) = C_0^\infty(0, \infty) \ ,$$

is to be decomposed at the point $x = 2$ and, for this purpose, we define the operator

$$
\begin{aligned}
&{}_+A_{2,0}u = {}_+A_0u \ , \\
&D({}_+A_{2,0}) = \{u \,|\, u \in C_0^\infty(0, \infty), \ u(x) = 0, \ |x-2| < \delta_u, \delta_u > 0\} \ .
\end{aligned}
$$

If the Friedrichs extensions of ${}_+A_0$ and ${}_+A_{2,0}$ are denoted by ${}_+\hat{A}$ and ${}_+\hat{A}_2$ respectively, then we have

$$\sigma_e({}_+\hat{A}) = \sigma_e({}_+\hat{A}_2) \tag{33}$$

as both operators are self-adjoint extensions of ${}_+A_{2,0}$. Now ${}_+\hat{A}_2$ is the orthogonal sum

$${}_+\hat{A}_2 = {}_+\hat{A}_{(0,2)} \oplus {}_+\hat{A}_{(2,\infty)}$$

of the operators ${}_+\hat{A}_{(0,2)}$ and ${}_+\hat{A}_{(2,\infty)}$, which arise from the operators

$${}_+A_{(2,\infty),0}u = \sum_{k=0}^{n^*} (-1)^k ({}_+a_k(x)\,u^{(k)})^{(k)}, \ D({}_+A_{(0,2),0}) = C_0^\infty(0, 2) \ ,$$

and $\qquad {}_+A_{(2,\infty),0}u = a[u], \ \ D({}_+A_{(2,\infty),0}) = C_0^\infty(2, \infty)$, by means of the Friedrichs extension. Hence

$$\sigma_e({}_+\hat{A}_2) = \sigma_e({}_+\hat{A}_{(0,2)}) \cup \sigma_e({}_+\hat{A}_{(2,\infty)})$$

by Theorem 1.7. We now prove that

$$\sigma_e({}_+\hat{A}_{(0,2)}) = \emptyset \ . \tag{34}$$

By Theorem 1.2, we do this by showing that every set bounded in the energy space $H_{+A(0,2),0}$ is pre-compact in $L_2(0, 2)$. With

$${}_+a_k(x) \geqslant 0,\ n+1 \leqslant k \leqslant n^*,\quad {}_+a_n(x) \geqslant \alpha > 0,\quad 0 < x < 2\ ,$$

and by (A 9), we obtain†

$$\sum_{k=0}^{n^*} \int_0^2 {}_+a_k(x)\,|u^{(k)}|^2\,\mathrm{d}x$$

$$\geqslant \alpha\,\|u^{(n)}\|^2_{(0,2)} - \sum_{k=0}^{n-1} \Big(\max_{0\leqslant x\leqslant 2} |u^{(k)}|^2\Big) \int_0^2 |{}_+a_k(x)|\,\mathrm{d}x$$

$$\geqslant \alpha\,\|u^{(n)}\|^2_{(0,2)} - \sum_{k=0}^{n-1} C_k\,(\epsilon\,\|u^{(n)}\|^2_{(0,2)} + C_\epsilon\,\|u\|^2_{(0,2)})\ ,$$

which gives

$$\sum_{k=0}^{n^*} \int_0^2 {}_+a_k(x)\,|u^{(k)}|^2\,\mathrm{d}x \geqslant \tfrac{1}{2}\alpha\,\|u^{(n)}\|^2_{(0,2)} - C_\epsilon\,\|u\|^2_{(0,2)}$$

if ϵ is sufficiently small. There are therefore positive constants c_1 and c_2 such that

$$\|u\|^2_{W_2^n(0,2)} \leqslant c_1 \sum_{k=0}^{n^*} \int_0^2 {}_+a_k(x)\,|u^{(k)}|^2\,\mathrm{d}x + c_2\,\|u\|^2_{(0,2)}\ , \tag{35}$$

$$u \in H_{+A(0,2),0}\ .$$

†$(\ .\ ,\ .\ )_{(x_1,x_2)}$ and $\|\cdot\|_{(x_1,x_2)}$, $-\infty \leqslant x_1 < x_2 \leqslant \infty$, denote the inner-product and norm in the Hilbert space $L_2(x_1,x_2)$.

Consequently each bounded set in $H_{+A_{(0,2),0}}$ is also bounded in $W_2^n(0, 2)$ and therefore pre-compact by Theorem A.2. Hence (34) is proved and we have

$$\sigma_e({}_+\hat{A}_2) = \sigma_e({}_+\hat{A}_{(2,\infty)}) \ . \tag{36}$$

Similarly A_0 is decomposed at the point $x = 2$. If $\hat{A}$, $\hat{A}_{(0,2)}$ and $\hat{A}_{(2,\infty)}$ are the Friedrichs extensions of the operators A_0,

$$A_{(0,2),0}u = a[u] \, , \quad D(A_{(0,2),0}) = C_0^\infty(0, 2) \ ,$$

and $$A_{(2,\infty),0}u = a[u] \, , \quad D(A_{(2,\infty),0}) = C_0^\infty(2, \infty) \ ,$$

respectively, then

$$\sigma_e(\hat{A}) = \sigma_e(\hat{A}_{(0,2)}) \cup \sigma_e(\hat{A}_{(2,\infty)}) \ .$$

There is again an estimate of the form (35) for $u \in H_{A_{(0,2),0}}$, and this results in

$$\sigma_e(\hat{A}_{(0,2)}) = \emptyset \ .$$

Consequently we have

$$\sigma_e(\hat{A}) = \sigma_e(\hat{A}_{(2,\infty)}) \ . \tag{37}$$

Since ${}_+\hat{A}_{(2,\infty)} = \hat{A}_{(2,\infty)}$, it follows from (33), (36) and (37) that

$$\sigma_e(\hat{A}) = \sigma_e({}_+\hat{A}) \ . \tag{38}$$

2) The functions ${}_+a_k(x)$ are now extended to the whole x-axis as even functions, that is,

$${}_+a_k(x) = {}_+a_k(-x), \quad -\infty < x < \infty, \quad k = 0, \ldots, n^* \ .$$

The Friedrichs extension ${}_*\hat{A}$ of

$${}_*A_0u = \sum_{k=0}^{n^*} (-1)^k \, ({}_+a_k(x) \, u^{(k)})^{(k)}, \quad D({}_*A_0) = C_0^\infty(-\infty, \infty) \ ,$$

is decomposed at $x = 0$. If the Friedrichs extension of

$${}_-A_0u = \sum_{k=0}^{n^*} (-1)^k \, ({}_+a_k(x) \, u^{(k)})^{(k)}, \quad D({}_-A_0) = C_0^\infty(-\infty, 0) \ ,$$

is denoted by $_{-}\hat{A}$, then we have

$$\sigma_e({}_*\hat{A}) = \sigma_e({}_-\hat{A}) \cup \sigma_e({}_+\hat{A}) \ . \tag{39}$$

Since $$\sigma_e({}_-\hat{A}) = \sigma_e({}_+\hat{A})$$

for reasons of symmetry about $x = 0$ it follows from (38) and (39) that

$$\sigma_e(A) = \sigma_e({}_*A) \ . \tag{40}$$

3) The operator

$${}_*B_0 u = \sum_{k=0}^{n} (-1)^k \, b_k u^{(2k)}, \quad D({}_*B_0) = C_0^\infty(-\infty, \infty) \ ,$$

is semibounded from below and essentially self-adjoint. We now show that the operators ${}_*A_0$ and ${}_*B_0$ (after adding where required a multiple of the identity operator) satisfy the conditions composed in Theorem 2.6 on the operators A_0 and B_0. First we prove the existence of positive constants c_1, c_2 such that

$$({}_*B_0 u, u) \leqslant c_1({}_*A_0 u, u) + c_2 \| u \|^2, \quad u \in C_0^\infty(-\infty, \infty)^\dagger .$$

Using (A.9), where ϵ is chosen sufficiently small, we obtain

$$\begin{aligned} ({}_*B_0 u, u) &= \sum_{k=0}^{n} \int_{-\infty}^{\infty} b_k \, | u^{(k)} |^2 \, dx \\ &\leqslant b_n \| u^{(n)} \|^2 + \epsilon \| u^{(n)} \|^2 + C_\epsilon \| u \|^2 \\ &\leqslant 2 b_n \| u^{(n)} \|^2 + C \| u \|^2 \ . \end{aligned} \tag{41}$$

To estimate $({}_*A_0 u, u)$ from below the properties

$$\inf_{-\infty < x < \infty} {}_+a_n(x) = \alpha, \quad {}_+a_k(x) \geqslant 0, \quad k = n+1, \ldots, n^* \ ,$$

$$\sup_{-\infty < x < \infty} \int_x^{x+1} | {}_+a_k(x) | \, dx \leqslant C_0, \quad k = 0, \ldots, n-1 \ ,$$

† Here $(. , .)$ and $\| \cdot \|$ denote the inner-product in the Hilbert space $L_2(-\infty, \infty)$.

and (A.9) are used. Then

$$({}_*A_0u, u) = \sum_{k=0}^{n^*} \int_{-\infty}^{\infty} {}_+a_k(x) \, | \, u^{(k)} \, |^2 \, dx$$

$$\geqslant \alpha \, \| \, u^{(n)} \, \|^2 - \sum_{k=0}^{n-1} \int_{-\infty}^{\infty} | \, {}_+a_k(x) \, \| \, u^{(k)} \, |^2 \, dx$$

$$= \alpha \, \| \, u^{(n)} \, \|^2 - \sum_{k=0}^{n-1} \sum_{\nu=-\infty}^{\infty} \int_{\nu}^{\nu+1} | \, {}_+a_k(x) \, | \, | \, u^{(k)} \, |^2 \, dx$$

$$\geqslant \alpha \, \| \, u^{(n)} \, \|^2 - \sum_{k=0}^{n-1} \sum_{\nu=-\infty}^{\infty} \left(\max_{x \in [\nu, \nu+1]} | \, u^{(k)} \, |^2 \right) \int_{\nu}^{\nu+1} | \, {}_+a_k(x) \, | \, dx$$

$$\geqslant \alpha \, \| \, u^{(n)} \, \|^2 - n \, C_0 \sum_{\nu=-\infty}^{\infty} (\epsilon \, \| \, u^{(n)} \, \|^2_{(\nu,\nu+1)} + C_\epsilon \, \| \, u \, \|^2_{(\nu,\nu+1)})$$

$$= \alpha \, \| \, u^{(n)} \, \|^2 - n \, C_0 \, \epsilon \, \| \, u^{(n)} \, \|^2 - n \, C_0 \, C_\epsilon \, \| \, u \, \|^2 \, .$$

With $\epsilon = \alpha/(2nC_0)$ in the last expression we obtain

$$({}_*A_0u, u) \geqslant \tfrac{1}{2}\alpha \, \| \, u^{(n)} \, \|^2 - C \, \| \, u \, \|^2, \quad u \in C_0^\infty(-\infty, \infty) \, . \tag{42}$$

Then (41) and (42) give

$$({}_*B_0u, u) \leqslant 4b_n\alpha^{-1} \, ({}_*A_0u, u) + C \, \| \, u \, \|^2, \quad u \in C_0^\infty(-\infty, \infty) \, .$$

Since ${}_*B_0$ is bounded from below, the coefficient b_0 can be modified so that

$$\| \, u \, \|^2 \leqslant ({}_*B_0u, u) \leqslant c({}_*A_0u, u), \quad c > 0, \quad u \in C_0^\infty(-\infty, \infty) \, , \tag{43}$$

and, in what follows, we assume this to have been done.

We now prove that the domain of the Friedrichs extension ${}_*\hat{B}$ of ${}_*B_0$ lies in $H_{{}_*A_0}$. Since ${}_*B_0$ is essentially self-adjoint by Theorem 2.4, $D({}_*\hat{B})$ arises from $C_0^\infty(-\infty, \infty)$ by completion in the norm

$$(\| {}_*B_0u \, \|^2 + \| \, u \, \|^2)^{\frac{1}{2}} \, .$$

This norm is equivalent to the norm of $W_2^{2n}(-\infty, \infty)$, so that

$$D({}_*\hat{B}) = W_2^{2n}(-\infty, \infty) .$$

Since also the estimate

$$\| u \|^2_{*A_0} = \sum_{k=0}^{n^*} \int_{-\infty}^{\infty} {}_+a_k(x) \, | u^{(k)} |^2 \, dx \leqslant C \| u \|^2_{W_2^{2n}(-\infty,\infty)} ,$$

$$u \in C_0^\infty(-\infty, \infty) ,$$

holds, $D({}_*B)$ is contained in H_{*A_0}.

We now choose a continuous positive function $p_0(x)$, $-\infty < x < \infty$, which tends to zero when $|x| \to \infty$, and we define

$$\eta_0(x) = | {}_+a_0(x) - b_0 | + p_0(x)$$

and

$$\eta_k(x) = \begin{cases} | {}_+a_k(x) - b_k |, & k = 1, \ldots, n , \\ {}_+a_k(x), & k = n + 1, \ldots, n^* . \end{cases}$$

In what follows we show that the form

$$s[u, v] = \sum_{k=0}^{n^*} \int_{-\infty}^{\infty} \eta_k(x) \, u^{(k)}(x) \bar{v}^{(k)}(x) \, dx , \quad D(s) = H_{*A_0} ,$$

satisfies the conditions in Theorem 2.6. Obviously (16) and (18) hold. It remains to prove that

$$s[u, u] \leqslant C \| u \|^2_{*A_0}, \quad u \in H_{*A_0} .$$

We have

$$s[u, u] \leqslant \sum_{k=n+1}^{n^*} \int_{-\infty}^{\infty} \eta_k(x) \, | u^{(k)} |^2 \, dx + \int_{-\infty}^{\infty} {}_+a_n(x) \, | u^{(n)} |^2 \, dx + C_1 \| u^{(n)} \|^2 + C_2 \|u\|^2 . \tag{44}$$

Also, choosing ϵ sufficiently small, we obtain

$$
\begin{aligned}
&({}_*\hat{A}u, u) \\
&\geqslant \sum_{k=n+1}^{n^*} \int_{-\infty}^{\infty} \eta_k \, |u^{(k)}|^2 \, dx + \int_{-\infty}^{\infty} {}_+a_n \, |u^{(n)}|^2 \, dx - \epsilon \, \| u^{(n)} \|^2 - C_\epsilon \| u \|^2 \\
&\geqslant \sum_{k=n+1}^{n^*} \int_{-\infty}^{\infty} \eta_k \, |u^{(k)}|^2 \, dx + \tfrac{1}{2} \int_{-\infty}^{\infty} {}_+a_n \, |u^{(n)}|^2 \, dx + (\tfrac{1}{2}\alpha - \varepsilon) \, \| u^{(n)} \|^2 - C_\epsilon \| u \|^2 \\
&\geqslant \sum_{k=n+1}^{n^*} \int_{-\infty}^{\infty} \eta_k \, |u^{(k)}|^2 \, dx + \tfrac{1}{2} \int_{-\infty}^{\infty} {}_+a_n \, |u^{(n)}|^2 \, dx + \tfrac{1}{3}\alpha \| u \|^2 - C \| u \|^2 .
\end{aligned}
\tag{45}
$$

Then from (44) and (45) it follows that

$$
s[u, u] \leqslant C_3 ({}_*\hat{A}u, u) + C_4 \| u \|^2 \leqslant C \| u \|^2_{{}_*A_0}, \quad u \in H_{{}_*A_0} .
$$

Finally we have to prove that the set ${}_*\hat{B}^{-1}M$ is s-pre-compact if the set M is bounded in $L_2(-\infty, \infty)$. From

$$
{}_*\hat{B}^{-1}h = u, \quad h \in M ,
$$

we obtain

$$
\sum_{k=1}^{n} (-1)^k \, b_k \, u^{(2k)} = h .
$$

Using (A.9) and choosing ϵ sufficiently small, we obtain

$$
\begin{aligned}
C_M &\geqslant \| \sum_{k=0}^{n-1} (-1)^k \, b_k \, u^{(2k)} \| \\
&\geqslant b_n \, \| u^{(2n)} \| - C_0 \sum_{k=0}^{n-1} \| u^{(2k)} \| \geqslant (b_n - \epsilon) \, \| u^{(2n)} \| - C_\epsilon \| u \| \\
&\geqslant \tfrac{1}{2} \, b_n \, \| u^{(2n)} \| - C \| u \| ,
\end{aligned}
$$

and hence

$$
\| u^{(2n)} \| \leqslant C_1 + C_2 \| u \| . \tag{46}
$$

Also, from $\| u \|^2 \leqslant ({}_*\hat{B}u, u) = (h, u) \leqslant \| h \| \, \| u \| \leqslant C_M \| u \|$, we obtain $$\|u\| \leqslant C_M, \quad u \in {}_*\hat{B}^{-1}M ,$$

which, together with (46), gives

$$\|u\|_{W_2^{2n}(-\infty,\infty)} \leqslant C, \quad u \in {}_*\hat{B}^{-1}M . \tag{47}$$

We have to prove that from any sequence $\{u_\nu\}_{\nu=1,2,\ldots}$, $u_\nu \in {}_*\hat{B}^{-1}M$, a subsequence $\{u_{\nu_j}\}_{j=1,2,\ldots}$ can be chosen such that

$$s[u_{\nu_j} - u_{\nu_{j'}}, \; u_{\nu_j} - u_{\nu_{j'}}] \to 0 \;, \quad j, j' \to \infty . \tag{48}$$

Let L be a positive integer chosen so that

$$\int_l^{l+1} \eta_k(x)\,\mathrm{d}x < \epsilon, \; l \text{ integer and } |l| \geqslant L, \quad 0 \leqslant k \leqslant n^* ,$$

for a given $\epsilon > 0$. This is possible by hypothesis 2) of the theorem. Then by (47),

$$\sum_{k=0}^{n^*} \int_{(-\infty,-L)\,\cup\,(L,\infty)} \eta_k(x)\,|u_\nu^{(k)} - u_\mu^{(k)}|^2\,\mathrm{d}x$$

$$\leqslant \sum_{k=0}^{n^*} \sum_{|l|\geqslant L} \int_l^{l+1} \eta_k(x)\,|u_\nu^{(k)} - u_\mu^{(k)}|^2\,\mathrm{d}x$$

$$\leqslant \sum_{k=0}^{n^*} \sum_{|l|\geqslant L} \Big(\max_{x\in[l,l+1]} |u_\nu^{(k)} - u_\mu^{(k)}|^2\Big) \int_l^{l+1} \eta_k(x)\,\mathrm{d}x \leqslant 2n\epsilon C^2$$

independently of μ and ν. Therefore, for a given $\epsilon > 0$, there exists a point $X_0(\epsilon)$ such that

$$\sum_{k=0}^{n^*} \int_{(-\infty,\,-X)\cup(X,\infty)} \eta_k(x)\,|u_\nu^{(k)} - u_\mu^{(k)}|^2\,\mathrm{d}x < \epsilon$$

for all $X \geqslant X_0(\epsilon)$ independently of μ and ν. Hence we obtain

$$s[u_\nu - u_\mu, u_\nu - u_\mu] \leqslant c \sum_{k=0}^{n^*} \int_{-X}^{X} |u_\nu^{(k)} - u_\mu^{(k)}|^2\,\mathrm{d}x + \epsilon, \quad X \geqslant X_0(\epsilon),$$

$$\mu, \nu = 1, 2, \ldots .$$

Since the embedding of $W_2^{2n}(-X, X)$ in $W_2^{n^*}(-X, X)$ is compact, there exists a subsequence $\{u_{\nu_j}\}_{j=1,2,\ldots}$ such that

$$s[u_{\nu_j} - u_{\nu_{j'}}, u_{\nu_j} - u_{\nu_{j'}}] \leqslant 2\epsilon, \quad j, j' \geqslant j_0(\epsilon).$$

Then the method of diagonal sequences yields the desired subsequence – it may again be denoted by $\{u_{\nu_j}\}_{j=1,2,\ldots}$ – for which (48) is true. Thus all conditions of Theorem 2.6 are satisfied, and it follows that

$$\sigma_e({}_*\hat{A}) = \sigma_e({}_*\hat{B}). \tag{49}$$

4) By Theorem 2.5 the essential spectrum of ${}_*\hat{B}$ is the interval

$$[\Lambda, \infty), \quad \Lambda = \inf_{-\infty<\xi<\infty} \sum_{k=0}^{n} b_k \xi^{2k}.$$

From (40) and (49) we now have

$$\sigma_e(\hat{A}) = [\Lambda, \infty).$$

Since all self-adjoint extensions of A_0 have the same essential spectrum, Theorem 2.9 is proved.

The above proof shows that

$$\sigma_e(\hat{A}) = [\Lambda, \infty)$$

still holds if the assumption that the highest coefficient $a_{n^*}(x)$ is positive for all x is dropped. It is sufficient to assume that $a_{n^*}(x) \geqslant 0$. However, the question then remains open whether all self-adjoint extensions of A_0 possess the same essential spectrum.

In what follows we consider perturbations of order $4n$ and, at the same time, weaken the assumption $a_k(x) \geqslant 0, k = n+1, \ldots, n^*$. The following theorem is an application of Theorem 2.8.

Theorem 2.10

Let the (real-valued) coefficients of the operator

$$A_0 u = \sum_{k=0}^{n^*} (-1)^k (a_k(x) u^{(k)})^{(k)}, \quad D(A_0) = C_0^\infty(0, \infty), \quad n \leqslant n^* \leqslant 2n,$$

satisfy the following conditions.

1) $\quad a_k(x) \in W_2^k(0, X)$ for all $X > 0, k = 0, \ldots, n^*$;

$$\inf_{0<x<\infty} a_n(x) = \alpha > 0, \quad a_{n^*}(x) > 0, \quad 0 < x < \infty.$$

2) For all $x \geqslant 0$ the estimate

$$\left| \sum_{k=n+1}^{n^*} \int_x^\infty a_k^-(t) \, | u^{(k)} |^2 \mathrm{d}t \right| \tag{50}$$

$$\leqslant \gamma \sum_{k=n+1}^{n^*} \int_x^\infty a_k^+(t) \, | u^{(k)} |^2 \, \mathrm{d}t + \alpha^* \, \| u^{(n)} \|_{(x,\infty)} + C^* \, \| u \|_{(x,\infty)},$$

$$a_k^-(x) = \min(a_k(x), 0), \quad a_k^+(x) = \max(a_k(x), 0), \quad u \in C_0^\infty(0, \infty),$$

holds, where the constants γ, α^* and C^* are independent of $u(x)$, and γ, α^* are restricted to $0 < \gamma < 1$ and $0 < \alpha^* < \alpha$ respectively.

3) There are constants b_k, $-\infty < b_k < \infty$, $k = 0, \ldots, n$, $b_n > 0$, such that

$$\lim_{x\to\infty} \int_x^{x+1} |a_k(t) - b_k| \, dt = 0, \quad 0 \leqslant k \leqslant n,$$

$$\lim_{x\to\infty} \int_x^{x+1} |a_k(t)| \, dt = 0, \quad n+1 \leqslant k \leqslant n^*.$$

If $n^* = 2n$, we also assume that

$$a_{2n}(x) \to 0, \; x \to \infty.$$

Then the essential spectrum of all self-adjoint extensions of A_0 is the interval

$$[\Lambda, \infty), \quad \Lambda = \inf_{0<\xi<\infty} \sum_{k=0}^{n} b_k \xi^{2k}.$$

Proof

The coefficients $a_k(x)$ are modified as follows. The functions

$$f(x) = \begin{cases} 1, & x \geqslant X + \frac{3}{2}, \quad X > 0, \\ 0, & x < X + \frac{3}{2}, \end{cases}$$

and

$$g(x) = \begin{cases} 1, & x \geqslant X + \frac{1}{2}, \\ 0, & x < X + \frac{1}{2}, \end{cases}$$

are smoothed with radius $\frac{1}{2}$ in each case. The smoothed functions $\phi(x)$ and $\psi(x)$ have the properties

$$\phi(x) \begin{cases} = 1, & x \geqslant X + 2, \\ = 0, & x \leqslant X + 1, \\ \in C^\infty(-\infty, \infty) \end{cases}$$

and

$$\psi(x)\begin{cases} = 1, & x \geqslant X+1, \\ = 0, & x \leqslant X, \\ \in C^\infty(-\infty, \infty) & \end{cases}$$

respectively. New coefficients are defined by

$$a_{k,X}(x) = \phi(x)\, a_k(x) + (1 - \phi(x)) b_k, \quad 0 \leqslant k \leqslant n,$$

$$a_{k,X}^{+}(x) = \psi(x)\, a_k^{+}(x), \quad n+1 \leqslant k \leqslant n^*,$$

$$a_{k,X}^{-}(x) = \phi(x) a_k^{-}(x), \quad n+1 \leqslant k \leqslant n^*,$$

$$a_{k,X}(x) = a_{k,X}^{+}(x) + a_{k,X}^{-}(x). \quad n+1 \leqslant k \leqslant n^*,$$

and the corresponding operator is

$$A_{X,0}u = \sum_{k=0}^{n^*} (-1)^k (a_{k,X}(x) u^{(k)})^{(k)}, \quad D(A_{X,0}) = C_0^\infty(0,\infty)\,.$$

We prove that this operator and B_0, where

$$B_0 u = \sum_{k=0}^{n} (-1)^k b_k u^{(2k)}, \quad D(B_0) = C_0^\infty(0,\infty),$$

form a pair which satisfies the conditions of Theorem 2.6. It is easy to see that condition (50) is also satisfied by the modified coefficients $a_{k,X}(x)$. Using this condition and also (A.9) with a sufficiently small ϵ, we obtain

$$(A_{X,0}u, u) \geqslant$$

$$\sum_{k=n+1}^{n^*} \int_0^\infty (a_{k,X}^{-}(x) + a_{k,X}^{+}(x)) \,|\, u^{(k)} \,|^2\, dx + \tfrac{1}{2}(\alpha - \alpha^*)\alpha^{-1} \int_0^\infty a_{n,X}(x) \,|\, u^{(n)} \,|^2\, dx +$$

$$\tfrac{1}{2}(\alpha + \alpha^*) \,\|\, u^{(n)} \,\|^2 - \epsilon \,\|\, u^{(n)} \,\|^2 - C_\epsilon \,\|\, u \,\|^2 \tag{51}$$

$$\geqslant (1-\gamma) \sum_{k=n+1}^{n^*} \int_0^\infty a_{k,X}^{+}(x) \,|\, u^{(k)} \,|^2\, dx + \tfrac{1}{2}(\alpha - \alpha^*) \int_0^\infty a_{n,X}(x) \,|\, u^{(n)} \,|^2\, dx +$$

$$\tfrac{1}{3}(\alpha - \alpha^*) \,\|\, u^{(n)} \,\|^2 - C \,\|\, u \,\|^2\,.$$

The coefficients of the form

$$s_X[u, v] = \sum_{k=0}^{n^*} \int_0^\infty \eta_{k,X}(x)\, u^{(k)}\, \bar{v}^{(k)} \mathrm{d}x, \quad u, v \in H_{A_{X,0}},$$

are defined by

$$\eta_{0,X}(x) = |a_{0,X}(x) - b_0| + (X + x)^{-1},$$

$$\eta_{k,X}(x) = |a_{k,X}(x) - b_k|, \quad 1 \leqslant k \leqslant n,$$

$$\eta_{k,X}(x) = |a_{k,X}(x)|, \qquad n + 1 \leqslant k \leqslant n^* .$$

Then we have

$$\begin{aligned} s_X[u, u] &= \sum_{k=0}^{n^*} \int_0^\infty \eta_{k,X}(x)\, | u^{(k)} |^2 \mathrm{d}x \\ &\leqslant \sum_{k=n+1}^{n^*} \int_0^\infty (a^+_{k,X}(x) - a^-_{k,X}(x))\, | u^{(k)} |^2 \mathrm{d}x + \int_0^\infty | a_{n,X}(x) |^2 \mathrm{d}x + \\ &\qquad C_1 \| u^{(n)} \|^2 + C_2 \| u \|^2 \qquad (52) \\ &\leqslant (1 + \gamma) \sum_{k=n+1}^{n^*} \int_0^\infty a^+_{k,X}(x)\, | u^{(k)} |^2 \mathrm{d}x + \int_0^\infty a_{n,X}(x)\, | u^{(n)} |^2 \mathrm{d}x + \\ &\qquad C_3 \| u^{(n)} \|^2 + C_4 \| u \|^2 . \end{aligned}$$

Lastly we have the estimate

$$(B_0 u, u) \leqslant 2b_n \| u^{(n)} \|^2 + C \| u \|^2, \quad u \in C_0^\infty(0, \infty). \qquad (53)$$

(cf. (41)). Now, (51) and (53) yield

$$(B_0 u, u) \leqslant 6b_n(\alpha - \alpha^*)^{-1} (A_{X,0} u, u) + C \| u \|^2, \quad u \in C_0^\infty(0, \infty), \qquad (54)$$

and from (51) and (52) we obtain

$$s_X[u, u] \leqslant \max\left(\frac{1 + \gamma}{1 - \gamma}, \frac{2\alpha}{\alpha - \alpha^*}, \frac{3\, C_3}{\alpha - \alpha^*}\right) \cdot (A_{X,0} u, u) + C \| u \|^2 . \qquad (55)$$

From (54) and (55) it follows in the first place that the coefficients b_0 and $a_0(x)$ can be chosen so that the inequalities

$$\| u \|^2 \leqslant (B_0 u, u) \leqslant C_1 (A_{X,0} u, u), \quad u \in C_0^\infty(0, \infty) \tag{56}$$

and

$$s_X [u, u] \leqslant C_2 (A_{X,0} u, u), \quad u \in C_0^\infty(0, \infty), \tag{57}$$

are valid, where the constants C_1 and C_2 are independent of X. Therefore the conditions (15)–(18) are satisfied by the operators $A_{X,0}$ and B_0.

Next we prove that $D(\hat{B}) \subseteqq H_{A_{X,0}}$. Let $\hat{u}(x)$ be any function in

$$D(\hat{B}) = D(B_0^*) \cap H_{B_0}.$$

We consider

$$\hat{u}_1(x) = \phi(x) \hat{u}(x), \quad 0 < x < \infty,$$

and

$$\hat{u}_2(x) = (1 - \phi(x)) \hat{u}(x), \quad 0 < x < \infty,$$

where $\phi(x)$ is as at the beginning of the proof. Since

$$D(B_0^*) = W_2^{2n}(0, \infty),$$

$\hat{u}_1(x)$ has the properties

$$\hat{u}_1(x) \begin{cases} \in W_2^{2n}(0, \infty), \\ = 0, \quad 0 < x < X + 1. \end{cases}$$

Therefore, since

$$(\hat{A}_X u, u) \leqslant C \| u \|^2_{W_2^{2n}(0,\infty)}, \quad u \in H_{A_{X,0}},$$

$\hat{u}_1(x)$ can be approximated by C_0^∞ functions in $H_{A_{X,0}}$. Hence

$$\hat{u}_1(x) \in H_{A_{X,0}}.$$

By Theorem 2.4 $\hat{u}_2(x)$ has the properties

$$\hat{u}_2(x) \in W_2^{2n}(0, \infty), \quad \hat{u}_2^{(\nu)}(0) = 0, \; 0 \leqslant \nu \leqslant n - 1, \quad \hat{u}_2(x) = 0, \; X + 2 < x < \infty,$$

and $\hat{u}_2(x)$ can likewise be approximated in $H_{A\,X,0}$ by C_0^∞ functions since on the interval $[0, X)$ the coefficients $a_{k,X}(x)$, $n+1 \leqslant k \leqslant n^*$, are identically zero. Then, since

$$\hat{u}(x) = \hat{u}_1(x) + \hat{u}_2(x)\,,$$

$\hat{u}(x)$ also belongs to $H_{A\,X,0}$, as we had to prove.

We can therefore define the operator T_X which corresponds to the operator T in the proof of Theorem 2.6. Thus we have

$$T_X\hat{B}^{-1} = \hat{B}^{-1} - \hat{A}_X^{-1}\,.$$

To apply Theorem 2.8 we estimate $\| \hat{B}^{-1} - \hat{A}_X^{-1} \|$. Let $h \in L_2(0, \infty)$ by any element with $\| h \| = 1$. Clearly the estimate

$$\| u \|_{W_2^{2n}(0,\infty)} \leqslant C, \qquad u = \hat{B}^{-1}h\,, \tag{58}$$

which is analogous to (47), holds. Next we use the estimate which is analogous to (26). We have

$$\| (\hat{B}^{-1} - \hat{A}_X^{-1})h \|^2 = \| T_X\hat{B}^{-1}h \|^2 = \| T_X u \|^2$$

$$\leqslant C \| T_X u \|^2_{A\,X,0} \leqslant C s_X[u, u] = C \sum_{k=0}^{n^*} \int_0^\infty \eta_{k,X}(x)\,|\,u^{(k)}\,|^2\,dx\,.$$

If we denote by (ν) the interval

$$(\nu) = \{x \mid X + \nu \leqslant x \leqslant X + \nu + 1\},\ \nu = 0, 1, \ldots,$$

and use (A.9), we obtain

$$\| (\hat{B}^{-1} - \hat{A}_X^{-1})h \|^2 \leqslant C \sum_{k=0}^{n^*} \sum_{\nu=0}^{\infty} \int_{(\nu)} \eta_{k,X}(x)\,|\,u^{(k)}\,|^2\,dx$$

$$\leqslant C \sum_{k=0}^{n^*-1} \sum_{\nu=0}^{\infty} \left(\int_{(\nu)} \eta_{k,X}(x)dx \right) \| u \|^2_{W_2^{2n}(\nu)} + C \sum_{\nu=0}^{\infty} F_{n^*}(\nu)\, \| u \|^2_{W_2^{2n}(\nu)}\,, \tag{59}$$

where

$$F_{n^*}(\nu) = \begin{cases} \displaystyle\int_{(\nu)} |\,a_{n^*,X}(x)\,|\,dx, & n^* < 2n\,, \\ \displaystyle\max_{x\in(\nu)} a_{2n}(x), & n^* = 2n\,. \end{cases}$$

By condition 3) of the theorem, for any given $\epsilon > 0$, there is a value $X = X(\epsilon)$ such that

$$\int_{(\nu)} \eta_{k,X}(x)\mathrm{d}x < \epsilon, \quad 0 \leqslant k \leqslant n^* - 1, \quad F_{n^*}(\nu) < \epsilon, \quad \nu = 0, 1, \ldots .$$

Hence from (59) we obtain

$$\| (\hat{B}^{-1} - \hat{A}_X^{-1}) h \|^2 \leqslant \epsilon\, C \| u \|^2_{W_2^{2n}(X,\infty)}$$

which, by (58), yields

$$\| (\hat{B}^{-1} - \hat{A}_X^{-1}) h \|^2 \leqslant \epsilon\, C, \quad \| h \| = 1 , \tag{60}$$

where C is independent of h and X. By (60), for a given positive integer μ, there is $X = X_\mu$ such that

$$\| \hat{B}^{-1} - \hat{A}_{X_\mu}^{-1} \| \leqslant \mu^{-1} . \tag{61}$$

To apply Theorem 2.8 we prove finally that

$$\sigma_e(\hat{A}_X) = \sigma_e(\hat{A}), \quad X > 0 . \tag{62}$$

To do this we decompose the operators at a point $X' \geqslant X + 2$. If $\hat{A}_{(X',\infty)}$ and $\hat{A}_{(0,X')}$ denote the Friedrichs extensions of

$$A_{(X',\infty),0}\, u = a[u], \; D(A_{(X',\infty),0}) = C_0^\infty(X', \infty) ,$$

and

$$A_{(0,X'),0} u = a[u], \; D(A_{(0,X'),0}) = C_0^\infty(0, X') ,$$

respectively, then we have

$$\sigma_e(\hat{A}) = \sigma_e(\hat{A}_{(0,X')}) \cup \sigma_e(\hat{A}_{(X',\infty)}) .$$

By (51), which is also valid for A_0 in place of $A_{0,X}$, we have

$$(A_{(0,X'),0} u, u)_{(0,X')} \geqslant \tfrac{1}{3}(\alpha - \alpha^*) \| u^{(n)} \|^2_{(0,X')} - C \| u \|^2_{(0,X')} ,$$

$$u \in C_0^\infty(0, X') . \tag{63}$$

Therefore a bounded set in $H_{A_{(0,X),0}}$ is also bounded in $W_2^n(0, X')$ and hence it is pre-compact in $L_2(0, X)$. Consequently, by Theorem 1.2, we have

$$\sigma_e(\hat{A}_{(0,X')}) = \emptyset,$$

and

$$\sigma_e(\hat{A}) = \sigma_e(\hat{A}_{(X',\infty)}) . \tag{64}$$

Similarly one can prove that

$$\sigma_e(\hat{A}_X) = \sigma_e(\hat{A}_{X,(X'\infty)}) , \tag{65}$$

where $\hat{A}_{X,(X',\infty)}$ denotes the Friedrichs extension for the expression

$$\sum_{k=0}^{n^*} (-1)^k (a_{k,X}(x)u^{(k)})^{(k)}$$

with domain $C_0^\infty(X',\infty)$. Since $\hat{A}_{(X',\infty)} = \hat{A}_{X,(X',\infty)}$, from (64) and (65) we have

$$\sigma_e(\hat{A}) = \sigma_e(\hat{A}_X) .$$

Therefore all operators $\hat{A}_{X_\mu}$, $\mu = 1, 2, \ldots$, possess the same essential spectrum. By Theorem 2.8 it now follows that

$$\sigma_e(\hat{A}) = \sigma_e(\hat{B}) ,$$

and Theorem 2.10 is proved.

Note 2.11

The condition (50) is satisfied if, for instance, there is a $C_0 > 0$ such that

$$\int_x^{x+1} |a_k^-(t)|\, dt \leqslant C_0 \cdot \min_{t \in [X,X+1]} a_{n^*}(t) , \quad n+1 \leqslant k \leqslant n^* - 1 , \tag{66}$$

for all $x \in (0, \infty)$.

Proof

If we denote

$$(\nu) = \{t \mid x + \nu \leqslant t \leqslant x + \nu + 1\}, \quad \nu = 0, 1, \ldots,$$

then

$$\int_{(\nu)} |a_k^-(t)| \, |u^{(k)}|^2 \, dt \leqslant \left(\max_{t \in (\nu)} |u^{(k)}(t)|^2 \right) \int_{(\nu)} |a_k^-(t)| \, dt$$

$$\leqslant C_0 \left(\min_{t \in (\nu)} a_{n^*}(t) \right) (\epsilon \, \| u^{(n^*)} \|^2_{(\nu)} + C_\epsilon \, \| u \|^2_{(\nu)})$$

$$\leqslant \epsilon C_0 \int_{(\nu)} a_{n^*}(t) \, | u^{(n^*)} |^2 \, dt + C_\epsilon \, \| u \|^2_{(\nu)} , \quad n+1 \leqslant k \leqslant n^* - 1 .$$

By summation over ν we obtain

$$\int_x^\infty |a_k^-(t)| \, |u^{(k)}|^2 \, dt \leqslant \epsilon C_0 \int_x^\infty a_{n^*}(t) \, | u^{(n^*)} |^2 \, dt + C_\epsilon \int_x^\infty |u|^2 \, dt ,$$

which yields

$$\left| \sum_{k=n+1}^{n^*} \int_x^\infty a_k^-(t) \, | u^{(k)} |^2 \, dt \right| \leqslant \epsilon \, n \, C_0 \int_x^\infty a_{n^*}(t) \, | u^{(n^*)} |^2 \, dt + nC_\epsilon \int_x^\infty |u|^2 \, dt$$

by summation over k. If ϵ is now chosen sufficiently small so that

$$\epsilon n C_0 = \gamma < 1 ,$$

then a particular case of (50) arises.

RELATIVELY COMPACT PERTURBATIONS

In what follows we shall further weaken part of the hypothesis of Theorem 2.10. To do this we use theorems of Birman on relatively compact perturbations. Let $b[u, v], D(b)$, be a positive definite and closed form:

$$b[u, u] \geqslant c \, \| u \|^2 , \quad c > 0 , \quad u \in D(b) . \tag{67}$$

By Theorem 1.5 the form $b[u, v]$ defines a positive self-adjoint operator B. Let $s[u, v]$ be a symmetric form, with $D(s) \supseteqq H_B$, which is bounded in H_B:

$$| s[u, v] | \leqslant C \, \| u \|_B \, \| v \|_B , \quad u, v \in H_B .$$

Then there is a self-adjoint bounded operator V, defined on H_B, such that

$$s[u, v] = [Vu, v]_B , \quad u, v \in H_B .$$

Definition 2.12
The form $s[u, v]$ is said to be relatively compact with respect to the form $b[u, v]$ if V is a compact operator in H_B.

If $s[u, v]$ is relatively compact with respect to $b[u, v]$, then certain properties of $b[u, v]$ are transferred to the form

$$a[u, v] = b[u,v] + s[u, v], \quad D(a) = D(b).$$

In this connexion we have

Theorem 2.13
Let the form $s[u, v]$ be relatively compact with respect to $b[u, v]$. Then for the form

$$a[u, v] = b[u, v] + s[u, v], \quad D(a) = D(b),$$

the following holds.

1) $a[u,v]$ is semibounded from below and closed; $\| \cdot \|_B$ is equivalent to the norm $\| \cdot \|_A$ defined by $a[u, v]$.

2) If A and B are self-adjoint operators defined by $a[u, v]$ and $b[u, v]$ respectively, then

$$\sigma_e(A) = \sigma_e(B).$$

Proof†

1) Let α_j, $j = 1, 2, \ldots$, be the non-zero eigenvalues of the self-adjoint operator V and let u_j, $j = 1, 2, \ldots$, be the corresponding eigenvectors, orthonormal in H_B. Since V is compact, $\alpha_j \to 0$ as $j \to \infty$, and

$$Vu = \sum_{j=1}^{\infty} \alpha_j [u, u_j]_B u_j$$

for any $u \in H_B$. Then for any $\epsilon > 0$ there is a number $j_0(\epsilon)$, such that $|\alpha_j| < \epsilon$ for $j > j_0(\epsilon)$, and we have

$$\| Vu \|_B^2 \leqslant \epsilon^2 \| u \|_B^2 + j_0 \| V \|^2 \cdot \max_{1 \leqslant j \leqslant j_0} | [u, u_j]_B |^2. \tag{68}$$

Since $D(B)$ is dense in H_B, for each $\eta > 0$ there are elements $\tilde{u}_j \in D(B)$ such that

$$\| u_j - \tilde{u}_j \|_B < \eta, \quad j = 1, \ldots, j_0.$$

† The proofs of the Theorems 2.13 and 2.14 are taken from transcripts of the lectures [4] and [55].

Hence

$$
\begin{aligned}
|[u, u_j]_B|^2 &= |[u, u_j - \tilde{u}_j]_B + [u, \tilde{u}_j]_B|^2 \\
&\leqslant 2|[u, u_j - \tilde{u}_j]_B|^2 + 2|[u, \tilde{u}_j]_B|^2 \\
&\leqslant 2\eta^2 \|u\|_B^2 + 2|(B\tilde{u}_j, u)|^2 \\
&\leqq 2\eta^2 \|u\|_B^2 + 2\|u\|^2 \|B\tilde{u}_j\|^2 .
\end{aligned}
\tag{69}
$$

If η is chosen so small that

$$2\eta^2 j_0 \|V\|^2 \leqslant \epsilon^2$$

then from (68) and (69) we obtain

$$\|Vu\|_B^2 \leqslant 2\epsilon^2 \|u\|_B^2 + C_\epsilon \|u\|^2 .$$

Hence for any $\epsilon' > 0$ there exists $C_{\epsilon'}$ such that

$$\|Vu\|_B \leqslant \epsilon' \|u\|_B + C_{\epsilon'} \|u\| ,$$

and this is used in the following estimate.

$$
\begin{aligned}
|s[u, u]| = |[Vu, u]_B| &\leqslant \epsilon' \|u\|_B^2 + C_{\epsilon'} \|u\| \|u\|_B \\
&\leqslant \epsilon' \|u\|_B^2 + \tfrac{1}{2} C_{\epsilon'} (\delta^2 \|u\|_B^2 + \delta^{-2} \|u\|^2) \\
&= (\epsilon' + \tfrac{1}{2} C_{\epsilon'} \delta^2) \|u\|_B^2 + \tfrac{1}{2} C_{\epsilon'} \delta^{-2} \|u\|^2 .
\end{aligned}
$$

Hence for any given $\epsilon > 0$ we can choose ϵ' and δ such that

$$|s[u, u]| \leqslant \epsilon \|u\|_B^2 + C_\epsilon \|u\|^2 , \quad u \in H_B ,$$

It follows finally that

$$a[u, u] = b[u, u] + s[u, u] \geqslant (1 - \epsilon) \|u\|_B^2 - C_\epsilon \|u\|^2 , \quad u \in H_B ,$$

and thus the semiboundedness of $a[u, u]$ is proved. Hence also the form $a[u, v] + c(u, v)$ is positive definite if c is chosen sufficiently large. Then the norm defined by

$$\|u\|_A = (a[u, u] + c\|u\|^2)^{\frac{1}{2}}$$

is equivalent to the norm $\|u\|_B$, as can be seen from the last two inequalities. It follows that $a[u, v]$ is closed on H_B.

2) Without loss of generality we now assume that the forms $a[u, v]$ and $b[u, v]$ are positive definite. Then the inverse operators A^{-1} and B^{-1} exist. The assertion $\sigma_e(A) = \sigma_e(B)$ is proved when the compactness of $B^{-1} - A^{-1}$ has been shown. To do this we consider the equations $Bu = h$ and $Av = h$ for any $h \in H$. With $g \in H_B$ we have

$$\begin{aligned} b[u, g] &= (Bu, g) = (Av, g) = a[v, g] \\ &= b[v, g] + s[v, g] = b[v, g] + [Vv, g]_B \\ &= b[v, g] + b[Vv, g] = b[v + Vv, g] \ , \end{aligned}$$

from which we obtain $u = v + Vv$, that is,

$$VA^{-1}h = (B^{-1} - A^{-1})h \ .$$

We prove that VA^{-1} is compact in H. The operator VA^{-1} is decomposed into four factors

$$VA^{-1} = E_{H_B \to H} \, V_{H_B \to H_B} \, A^{-\frac{1}{2}}_{H \to H_B} \, A^{-\frac{1}{2}}_{H \to H} \ ,$$

where $A^{-\frac{1}{2}}_{H \to H}$ transforms H into itself and is continuous, and $A^{-\frac{1}{2}}_{H \to H_B}$ is a transformation from H to H_B in accordance with the fact that the domain of $A^{-\frac{1}{2}}$ is $H_B (= H_A)$ – see Theorem 1.5, part (4). Since

$$\begin{aligned} &\| A^{-\frac{1}{2}} h \|_A^2 = a[A^{-\frac{1}{2}} h, A^{-\frac{1}{2}} h] \\ &= (A^{\frac{1}{2}} A^{-\frac{1}{2}} h, A^{\frac{1}{2}} A^{-\frac{1}{2}} h) = \| h \|^2 \ , \quad h \in H \ , \end{aligned}$$

and in view of the equivalence of the norms $\| \cdot \|_A$ and $\| \cdot \|_B$, it follows that $A^{-\frac{1}{2}}_{H \to H_B}$ is also a continuous operator. Now, $V = V_{H_B \to H_B}$ is compact and the embedding $E_{H_B \to H}$ is continuous since

$$\| u \|_B \geqslant C \| u \| \ , C > 0 \ , u \in H_B \ .$$

Hence, altogether, the operator

$$VA^{-1} = B^{-1} - A^{-1}$$

is compact and Theorem 2.13 is proved.

The next theorem deals with the question of when a form is relatively compact with respect to a given positive definite form.

Theorem 2.14

Let $b[u, v]$, $D(b)$, be a positive definite, closed form and let B be the self-adjoint operator defined by $b[u, v]$. Then the symmetric form $s[u, v]$, $D(s) \supseteq H_B$, is relatively compact with respect to $b[u, v]$ if the following is valid.

1) $|s[u,u]| \leqslant C\|u\|_B^2\,, \quad C>0\,, \quad u \in H_B\,.$

2) In each infinite set M bounded in H_B there is a sequence $\{u_j\}_{j=1,2,\ldots}$ such that

$$s[u_j - u_{j'}, u_j - u_{j'}] \to 0, \quad j, j' \to \infty\,.$$

Proof

By 1) $s[u,v]$ is bounded in H_B. Therefore there is a bounded self-adjoint operator V in H_B such that

$$s[u,v] = [Vu, v]_B, \quad u, v \in H_B\,,$$

and we have to prove that V is compact. For this purpose V is decomposed according to

$$V = \int_{-\|V\|}^{\|V\|} \lambda \mathrm{d}E_\lambda = \int_{-\|V\|}^{0} \lambda \mathrm{d}E_\lambda + \int_{0}^{\|V\|} \lambda \mathrm{d}E_\lambda\,.$$

We show that each of the operators

$$V^- = \int_{-\|V\|}^{0} \lambda \mathrm{d}E_\lambda \quad \text{and} \quad V^+ = \int_{0}^{\|V\|} \lambda \mathrm{d}E_\lambda$$

is compact. With the projections E_0 and $E - E_0$, the mutually orthogonal subspaces

$$H_B^- = E_0 H_B \quad \text{and} \quad H_B^+ = (E - E_0) H_B$$

are defined. By hypothesis 2) we can select a sequence $\{u_j^+\}_{j=1,2\ldots.}$ from a bounded set in H_B^+ such that

$$s[u_j^+ - u_k^+, u_j^+ - u_k^+] \to 0, \quad j, k \to \infty\,,$$

and similarly we can select a sequence $\{u_j^-\}_{j=1,2,\ldots}$ from a bounded set in H_B^- such that

$$s[u_j^- - u_k^-, u_j^+ - u_k^+] \to 0, \quad j, k \to \infty\,.$$

Then, from

$$\begin{aligned}\|(V^+)^{\frac{1}{2}}(u_j^+ - u_k^+)\|_B^2 &= [V^+(u_j^+ - u_k^+), u_j^+ - u_k^+]_B \\ &= [V(u_j^+ - u_k^+), u_j^+ - u_k^+]_B \\ &= s[u_j^+ - u_k^+, u_j^+ - u_k^+] \to 0, \quad j, k \to \infty\,,\end{aligned}$$

it follows that $(V^+)^{\frac{1}{2}}$ is compact in H_B^+. Hence V^+ is also compace in H_B^+. similarly

$$-V^- = \int_{-\|V\|}^{0} |\lambda| \, dE_\lambda$$

is compact in H_B^-, and it follows that $V = V^- + V^+$ is compact in $H_B = H_B^- \oplus H_B^+$. This proves Theorem 2.14.

To prepare for the application of Theorem 2.14 to differential operators we prove the following lemma.

Lemma 2.15 [37]
Let $p(x)$ and $q(x)$ be real-valued functions defined on the interval $\omega = [x_1, x_2]$, with length $|\omega|$, where $p(x) \geqslant 0$ and $p^{-1}(x)$ and $q(x)$ are integrable over ω. Let

$$\mu_\omega = \frac{1}{|\omega|} \int_\omega q(x) \, dx$$

denote the mean value of $q(x)$ on ω, and let either $q(x) \geqslant 0$ or $q(x) \leqslant 0$ on ω. Then

$$\begin{aligned} &\int_\omega (p(x)|u'|^2 + q(x)|u|^2) \, dx \\ &\geqslant \mu_\omega \left(1 + |\omega| \mu_\omega \int_\omega p^{-1}(x) \, dx\right)^{-1} \|u\|_\omega^2 \end{aligned} \tag{70}$$

for every (complex-valued) function $u(x) \in C^1[x_1, x_2]$; in the second case $|\omega|$ must be chosen so small that

$$\left(\int_\omega |q(x)| \, dx\right) \left(\int_\omega p^{-1}(x) \, dx\right) < 1 .$$

Proof
Let $\rho(x) \in C^1[x_1, x_2]$ be a real-valued function with a zero x_0 in $[x_1, x_2]$. Then, by the Schwarz inequality, we obtain

$$\begin{aligned} \rho^2(x) &= \left(\int_{x_0}^{x} \rho'(t) \, dt\right)^2 \leqslant \left(\int_{x_0}^{x} p^{-1}(t) \, dt\right) \left(\int_{x_0}^{x} p(t)[\rho'(t)]^2 \, dt\right) \\ &\leqslant \left(\int_\omega p^{-1}(t) \, dt\right) \left(\int_\omega p(t)[\rho'(t)]^2 \, dt\right), \end{aligned}$$

and hence, by integration,

$$\|\rho\|_\omega^2 \leqslant |\omega| \left(\int_\omega p^{-1}(x)\,\mathrm{d}x\right)\left(\int_\omega p(x)[\rho'(x)]^2\,\mathrm{d}x\right). \tag{71}$$

Let $u_0 = \varphi_0 + i\psi_0$ denote such a value of the complex-valued function

$$u(x) = \varphi(x) + i\psi(x)$$

that $|u(x)|$ is minimal on ω. Then from (71) we obtain

$$\int_\omega (\varphi(x) - \varphi_0)^2\,\mathrm{d}x \leqslant |\omega| \left(\int_\omega p^{-1}(x)\,\mathrm{d}x\right)\left(\int_\omega p(x)[\varphi'(x)]^2\,\mathrm{d}x\right)$$

and

$$\int_\omega (\psi(x) - \psi_0)^2\,\mathrm{d}x \leqslant |\omega| \left(\int_\omega p^{-1}(x)\,\mathrm{d}x\right)\left(\int_\omega p(x)[\psi'(x)]^2\,\mathrm{d}x\right).$$

By addition we get

$$\int_\omega |u(x) - u_0|^2\,\mathrm{d}x \leqslant |\omega| \left(\int_\omega p^{-1}(x)\,\mathrm{d}x\right)\left(\int_\omega p(x)\,|u'(x)|^2\,\mathrm{d}x\right). \tag{72}$$

We consider first the case $q(x) \geqslant 0, x \in \omega$. Then, by (72),

$$\int_\omega (p(x)\,|u'|^2 + q(x)\,|u|^2)\,\mathrm{d}x$$

$$\geqslant \left(|\omega| \int_\omega p^{-1}(x)\,\mathrm{d}x\right)^{-1} \|u - u_0\|_\omega^2 + \mu_\omega \|u_0\|_\omega^2$$

$$= \frac{\mu_\omega}{1+\delta^2}\left[\delta^2\left(|\omega|\,\mu_\omega \int_\omega p^{-1}(x)\,\mathrm{d}x\right)^{-1}(1+\delta^{-2})\,\|u - u_0\|_\omega^2 + (1+\delta^2)\,\|u_0\|_\omega^2\right.$$

for any $\delta > 0$. If we choose

$$\delta^2 = |\omega|\,\mu_\omega \int_\omega p^{-1}(x)\,\mathrm{d}x$$

and use the inequality

$$\| u \|_\omega^2 \leqslant (1 + \delta^{-2}) \| u - u_0 \|_\omega^2 + (1 + \delta^2) \| u_0 \|_\omega^2 ,$$

then we obtain (70).

Now suppose that $q(x) \leqslant 0, x \in \omega$. Now (72) is also valid if u_0 is such a value of $u(x)$ for which $| u(x) |$ is a maximum . With this meaning for u_0, we have

$$\begin{aligned} \| u_0 \|_\omega^2 &\leqslant (\| u_0 - u \|_\omega + \| u \|_\omega)^2 \\ &\leqslant \| u \|_\omega^2 + | \omega | \left(\int_\omega p^{-1}(x)\,dx \right) \left(\int_\omega p(x) \, | u' |^2 \,dx \right) \\ &\qquad + 2 \| u \|_\omega \, | \omega |^{\frac{1}{2}} \left(\int_\omega p^{-1}(x)\,dx \right)^{\frac{1}{2}} \left(\int_\omega p(x) \, | u' |^2 \,dx \right)^{\frac{1}{2}} \\ &\leqslant (1 + \delta^2) \| u \|_\omega^2 + (1 + \delta^{-2}) | \omega | \left(\int_\omega p^{-1}(x)\,dx \right) \left(\int_\omega p(x) | u' |^2 \,dx \right) . \end{aligned}$$

Hence

$$\begin{aligned} \int_\omega \left(p(x) | u' |^2 + q(x) | u |^2 \right) dx &\geqslant \int_\omega p(x) \, | u' |^2 \,dx + \mu_\omega \| u_0 \|_\omega^2 \\ \geqslant \left[1 + \mu_\omega (1 + \delta^{-2}) \, | \omega | \left(\int_\omega p^{-1}(x)\,dx \right) \right] &\cdot \int_\omega p(x) \, | u' |^2 \,dx \\ &+ \mu_\omega (1 + \delta^2) \| u \|_\omega^2 . \end{aligned}$$

Since in the case $\mu_\omega < 0$, we can choose

$$\delta^2 = - \mu_\omega \, | \omega | \left(\int_\omega p^{-1}(x)\,dx \right) \Big/ \left(1 + \mu_\omega \, |\omega| \int_\omega p^{-1}(x)\,dx \right) ,$$

provided that $| \omega |$ is so small that

$$\left(\int_\omega | q(x) | \,dx \right) \left(\int_\omega p^{-1}(x)\,dx \right) < 1 ,$$

(70) follows again and Lemma 2.15 is proved.

If $q(x)$ takes both signs on ω, we use the decomposition

$$q(x) = q^-(x) + q^+(x), \quad q^-(x) = \min(q(x), 0), \quad q^+(x) = \max(q(x), 0) ,$$

and, with the mean values

$$\mu_\omega(q^-) = \frac{1}{|\omega|}\int_\omega q^-(x)\,dx,\quad \mu_\omega(q^+) = \frac{1}{|\omega|}\int_\omega q^+(x)\,dx\,,$$

$$\mu_\omega(q) = \frac{1}{|\omega|}\int_\omega q(x)\,dx\,,$$

we obtain from (70)

$$\int_\omega (p(x)|u'|^2 + q(x)|u|^2)\,dx$$

$$= \int_\omega (\tfrac{1}{2}p(x)|u'|^2 + q^-(x)|u|^2)\,dx + \int_\omega (\tfrac{1}{2}p(x)|u'|^2 + q^+(x)|u|^2)\,dx$$

$$\geqslant \left(\frac{\mu_\omega(q^-)}{1+2|\omega|\mu_\omega(q^-)\int_\omega p^{-1}(x)\,dx} + \frac{\mu_\omega(q^+)}{1+2|\omega|\mu_\omega(q^+)\int_\omega p^{-1}(x)\,dx}\right)\|u\|_\omega^2$$

$$= \frac{\mu_\omega(q) + 4|\omega|\mu_\omega(q^-)\mu_\omega(q^+)\int_\omega p^{-1}(x)\,dx}{(1+2|\omega|\mu_\omega(q^-)\int_\omega p^{-1}(x)\,dx)(1+2|\omega|\mu_\omega(q^+)\int_\omega p^{-1}(x)\,dx)}\|u\|_\omega^2\,. \tag{73}$$

Here $|\omega|$ must be so small that

$$\left(\int_\omega |q^-(x)|\,dx\right)\left(\int_\omega p^{-1}(x)\,dx\right) < \tfrac{1}{2}\,.$$

To prove the next theorem we need (73) for the function $p(x) = h = \text{const}$. If we put $q(x) = -s(x)$ and use the mean values

$$\mu_\omega(s^-) = \frac{1}{|\omega|}\int_\omega s^-(x)\,dx,\quad \mu_\omega(s^+) = \frac{1}{|\omega|}\int_\omega s^+(x)\,dx,$$

$$\mu_\omega(s) = \frac{1}{|\omega|}\int_\omega s(x)\,dx\,,$$

we obtain

$$\int_\omega s(x)|u|^2\,dx$$

$$\leqslant h\|u'\|_\omega^2 + \frac{\mu_\omega(s) + 4h^{-1}|\omega|^2\,|\mu_\omega(s^-)\mu_\omega(s^+)|}{(1-2h^{-1}|\omega|^2\mu_\omega(s^-))(1-2h^{-1}|\omega|^2\mu_\omega(s^+))}\|u\|_\omega^2\,.$$

The corresponding inequality with $-s(x)$ in place of $s(x)$ gives

$$-\int_{\omega} s(x)|u|^2 \,dx$$

$$\leqslant h\|u'\|_{\omega}^2 - \frac{\mu_{\omega}(s) - 4h^{-1}|\omega|^2|\mu_{\omega}(s^-)\mu_{\omega}(s^+)|}{(1+2h^{-1}|\omega|^2\mu_{\omega}(s^-))(1+2h^{-1}|\omega|^2\mu_{\omega}(s^+))}\|u\|_{\omega}^2 .$$

Then from these two inequalities it follows that

$$\left|\int_{\omega} s(x)|u|^2 \,dx\right|$$

$$\leqslant h\|u'\|_{\omega}^2 + \frac{|\mu_{\omega}(s)| + 4h^{-1}|\omega|^2|\mu_{\omega}(s^-)\mu_{\omega}(s^+)|}{(1-2h^{-1}|\omega|^2|\mu_{\omega}(s^+)|)(1-2h^{-1}|\omega|^2|\mu_{\omega}(s^-)|)}\|u\|_{\omega}^2 \tag{74}$$

provided that

$$1 - 2h^{-1}|\omega|^2\mu_{\omega}(s^+) > 0,\ 1 - 2h^{-1}|\omega|^2|\mu_{\omega}(s^-)| > 0 .$$

Now we make use of Theorem 2.14.

Theorem 2.16 [38]
Let all the conditions of Theorem 2.10 about the coefficients of the operator

$$A_0 u = \sum_{k=0}^{n^*} (-1)^k (a_k(x) u^{(k)})^{(k)},\quad D(A_0) = C_0^{\infty}(0,\infty),$$

be satisfied except those which concern the coefficients $a_k(x)$, $0 \leqslant k \leqslant n-1$. The conditions for these are weakened as follows.

i) Let

$$a_k(x) \in W_2^k(0, X) \quad \text{for every} \quad X > 0,$$

and

$$\sup_{0<x<\infty} \int_x^{x+|\omega|} |a_k(t)|\,dt \to 0,\quad |\omega| \to 0,\quad 0 \leqslant k \leqslant n-1 .$$

ii) For each fixed $|\omega|>0$, let

$$\lim_{x\to\infty}\frac{1}{|\omega|}\int_x^{x+|\omega|} a_k(t)\,dt = b_k,\quad 0\leqslant k\leqslant n-1\,.$$

Then A_0 is semibounded from below and the essential spectrum of every self-adjoint extension of A_0 is the interval

$$[\Lambda,\infty),\quad \Lambda=\inf_{0<t<\infty}\sum_{k=0}^{n} b_k t^{2k}\,.$$

Proof
We define

$$\beta_k(x)=a_k(x),\quad 0<x<\infty,\quad n\leqslant k\leqslant n^*,$$
$$\beta_k(x)=b_k,\quad 0\leqslant k\leqslant n-1\,,$$

and consider the form

$$b_0[u,v] = \sum_{k=0}^{n^*}\int_0^\infty \beta_k(x)u^{(k)}(x)\bar{v}^{(k)}(x)\,dx,\quad D(b_0)=C_0^\infty(0,\infty)\,,$$

which is semibounded from below. We may again assume that $b_0[u,v]$ is positive definite. Let $b[u,v]$ be the closure of $b_0[u,v]$. By Theorem 2.10,

$$\sigma_e(B)=[\Lambda,\infty)\,,$$

where B is the self-adjoint operator defined by $b[u,v]$. We define

$$s[u,v] = \sum_{k=0}^{n-1}\int_0^\infty [a_k(x)-b_k]\,u^k\bar{v}^{(k)}dx,\quad u,v\in H_B\;,$$

and prove that $s[u,v]$ is relatively compact with respect to $b[u,v]$. We put $s_k(x)=a_k(x)-b_k$ and estimate the integral of $s_k(x)|\,u^{(k)}\,|^2$ using (74) with $s(x)$ replaced by $s_k(x)$ and $u(x)$ by $u_k(x)$. By condition (i), which also holds for $s_k(x)$, and

$$\text{ii)}\quad \lim_{x\to\infty}\frac{1}{|\omega|}\int_x^{x+|\omega|} s_k(t)\,dt=0,\quad 0\leqslant k\leqslant n-1\,,\quad \text{for each } |\omega|>0\,,$$

we obtain from (74) with a fixed $h > 0$ the estimate

$$\left| \int_x^{x+|\omega|} s_k(t) |u^{(k)}|^2 \, dt \right| \leqslant h \, \| u^{(k+1)} \|^2_{(x,x+|\omega|)} + h \, \| u^{(k)} \|^2_{(x,x+|\omega|)} , \tag{75}$$

$$|\omega| \leqslant |\omega_0(h)|, \quad x \geqslant x_0(h) ,$$

provided that $|\omega|$ is first chosen sufficiently small and then x sufficiently large. By joining together intervals of length $|\omega|$ and adding the corresponding inequalities (75) we obtain

$$\left| \int_x^{\infty} s_k(t) |u^{(k)}|^2 \, dt \right| \leqslant h(\| u^{(k+1)} \|^2_{(x,\infty)} + \| u^{(k)} \|^2_{(x,\infty)}) , \tag{76}$$

and hence

$$\left| \sum_{k=0}^{n-1} \int_x^{\infty} s_k(t) |u^{(k)}|^2 \, dt \right| \leqslant nh \, \| u \|^2_{W_2^n(0,\infty)}, \quad x \geqslant x_0(h) . \tag{77}$$

The inequality 1) of Theorem 2.14 now follows from

$$|s[u,u]| \leqslant C \, \| u \|^2_{W_2^n(0,\infty)}$$

and

$$\| u \|^2_{W_2^n(0,\infty)} \leqslant C b[u,u] ,$$

which in turn results from (51) if $a_{k,X}(x)$ is there replaced by β_k.

To prove that condition 2) of Theorem 2.14 is satisfied we consider a set M bounded in H_B. Let $\epsilon > 0$ be arbitrary. From (77) it follows that, in the inequality

$$\left| \sum_{k=0}^{n-1} \int_0^{\infty} s_k(x) |u_j^{(k)}(x) - u_{j'}^{(k)}(x)|^2 \, dx \right| \tag{78}$$

$$\leqslant \left| \sum_{k=0}^{n-1} \int_0^{x_1} s_k(x) |u_j^{(k)} - u_{j'}^{(k)}|^2 \, dx \right| + \left| \sum_{k=0}^{n-1} \int_{x_1}^{\infty} s_k(x) |u_j^{(k)} - u_{j'}^{(k)}|^2 \, dx \right| ,$$

the point $x_1 = x_1(\epsilon)$ can be chosen so that independently of $u_j, u_{j'} \in M, j, j' = 1, 2, \ldots$, the inequality

$$\left| \sum_{k=0}^{n-1} \int_{x_1}^{\infty} s_k(x) | u_j^{(k)} - u_{j'}^{(k)} |^2 \, dx \right| < \frac{\epsilon}{2}, \quad j, j' = 1, 2, \ldots,$$

holds. Since the embedding of $W_2^n(0, x_1)$ in $C^{n-1}[0, x_1]$ is compact, there exists a sequence $\{u_j\}_{j=1,2,\ldots}$, $u_j \in M$ – let this sequence be the only already used above – such that

$$\sum_{k=0}^{n-1} \max_{x \in [0, x_1]} | u_j^{(k)}(x) - u_{j'}^{(k)}(x) |^2 \to 0, \quad j, j' \to \infty .$$

Hence from (78) we have

$$\left| \sum_{k=0}^{n-1} \int_0^{\infty} s_k(x) | u_j^{(k)}(x) - u_{j'}^{(k)}(x) |^2 \, dx \right| < \epsilon, \quad j, j' > j_0(\epsilon) .$$

The method of diagonal sequences yields the desired subsequence of $\{u_j\}_{j=1,2,\ldots}$ – let us denote it by $\{u_j\}_{j=1,2,\ldots}$ again – for which

$$\sum_{k=0}^{n-1} \int_0^{\infty} s_k(x) | u_j^{(k)}(x) - u_{j'}^{(k)}(x) |^2 \, dx \to 0, \quad j, j' \to \infty .$$

This proves Theorem 2.16

Note 2.17

Theorem 2.16 refers to the operator A_0 with the basic interval $(0, \infty)$. To obtain the same result for the essential spectrum in the case of the basic interval $(-\infty, \infty)$, the conditions must be changed as follows.

1) $a_k(x) \in W_{2,\mathrm{loc}}^k(-\infty, \infty), \quad 0 \leqslant k \leqslant n^*; \quad \inf_{-\infty < x < \infty} a_n(x) = \alpha > 0 .$

$a_{n^*}(x) > 0, \quad -\infty < x < \infty, \quad n \leqslant n^* \leqslant 2n .$

2) For all $x \geqslant 0$ the estimates (50) and

$$\left| \sum_{k=n+1}^{n^*} \int_{-\infty}^{-x} a_k^-(t) | u^{(k)} |^2 \, dt \right|$$

$$\leqslant \sigma \sum_{k=n+1}^{n^*} \int_{-\infty}^{-x} a_k^+(t) | u^{(k)} |^2 \, dt + \alpha^* \| u^{(n)} \|_{(-\infty, -x)}^2 + C^* \| u \|_{(-\infty, -x)}^2$$

$$0 < \gamma < 1 , \quad 0 < \alpha^* < \alpha, \quad u \in C_0^{\infty}(-\infty, \infty) ,$$

hold.

3) i) $\sup_{-\infty<x<\infty} \int_x^{x+|\omega|} |a_k(t)| dt \to 0, \ |\omega| \to 0, \ 0 \leqslant k \leqslant n-1 .$

ii) For each fixed $|\omega| > 0$ we have

$$\lim_{|x|\to\infty} \frac{1}{|\omega|} \int_x^{x+|\omega|} |a_k(t) \, dt = b_k \,, \quad 0 \leqslant k \leqslant n-1 .$$

iii) $\lim_{|x|\to\infty} \int_x^{x+1} |a_n(t) - b_n| \, dt = 0 ,$

$$\lim_{|x|\to\infty} \int_x^{x+1} |a_k(t)| \, dt = 0, \quad n+1 \leqslant k \leqslant n^* .$$

iv) $\lim_{|x|\to\infty} a_{2n}(x) = 0$ when $n^* = 2n$.

Then all self-adjoint extensions of A_0 have the same essential spectrum

$$[\Lambda, \infty), \quad \Lambda = \inf_{-\infty<t<\infty} \sum_{k-0}^{n} b_k t^{2k} .$$

Other perturbations than those described in Theorems 2.9, 2.10, and 2.16 can also be made and, under suitable conditions, the essential spectrum remains unchanged again. We consider an example.

For every self-adjoint extension A of A_0,

$$A_0 u = \sum_{k=0}^{n} (-1)^k (a_k(x) u^{(k)})^{(k)}, \quad D(A_0) = C_0^\infty(0, \infty) ,$$

we have

$$\sigma_e(A) = [\Lambda, \infty), \quad \Lambda = \inf_{0<\xi<\infty} \sum_{k=0}^{n} b_k \xi^{2k} ,$$

if the following conditions are satisfied.

1) $a_k(x) \in W_2^k(0, X)$ for all $X > 0$, $0 \leqslant k \leqslant n$.

2) There exist real numbers b_k such that

$$a_k(x) \geqslant b_k, \quad 0 < x < \infty, \quad 0 \leqslant k \leqslant n, \quad b_n > 0 .$$

3) There are mutually disjoint intervals $[\xi_\nu - l_\nu, \xi_\nu + l_\nu]$, $\nu = 1, 2, \ldots$, in $(0, \infty)$, with $\xi_\nu \to \infty$ and $l_\nu \to \infty$ when $\nu \to \infty$, such that

$$a_k(x) = b_k, \ \xi_\nu - l_\nu \leqslant x \leqslant \xi_\nu + l_\nu, \ \nu = 1,2, \ldots, \quad 0 \leqslant k \leqslant n .$$

This fact can easily be proved by means of a Weyl sequence (see Theorem 2.5) and the Courant Variational Principle (Theorem 1.6). If now the coefficients $a_k(x)$, $0 \leqslant k \leqslant n-1$, are perturbed by functions $s_k(x)$ which satisfy the conditions

$$\text{i)} \qquad \sup_{0<x<\infty} \int_x^{x+|\omega|} |s_k(t)|\,dt \to 0, \quad |\omega| \to 0, \quad 0 \leqslant k \leqslant n-1\,,$$

$$\text{ii)} \qquad \lim_{x\to\infty} \frac{1}{|\omega|} \int_x^{x+|\omega|} s_k(t)\,dt = 0, \quad 0 \leqslant k \leqslant n-1\,,$$

$$\text{iii)} \qquad \inf_{0<x<\infty} [a_n(x) + s_n(x)] > 0\,,$$

then the essential spectrum of A remains unchanged. This is a simple consequence of the proof of Theorem 2.16.

EXISTENCE OF ISOLATED EIGENVALUES

In the foregoing theorems no mention has been made of whether there are eigenvalues of self adjoint extensions of A_0 below the lowest point Λ of the essential spectrum and whether there are only a finite or an infinite number of such eigenvalues. We prove two theorems concerning these questions in this section.

Theorem 2.18
Let the (real-valued) coefficients of the operator

$$A_0 u = \sum_{k=0}^{n^*} (-1)^k (a_k(x) u^{(k)})^{(k)}, \quad D(A_0) = C_0^\infty(-\infty, \infty)\,,$$

satisfy the following conditions.

1) $a_k(x) \in W^k_{2,\mathrm{loc}}(-\infty,\infty)$, $0 \leqslant k \leqslant n^*$, $n \leqslant n^* \leqslant 2n$;

$$\inf_{-\infty<x<\infty} a_n(x) = \alpha > 0, \quad a_{n^*}(x) > 0, \quad -\infty < x < \infty\,.$$

2) For all $x \geqslant 0$ the estimates (50) and

$$\left| \sum_{k=n+1}^{n^*} \int_{-\infty}^{-x} a_k^-(t) |u^{(k)}|^2\,dt \right| \tag{79}$$

$$\leqslant \gamma \sum_{k=n+1}^{n^*} \int_{-\infty}^{-x} a_k^+(t) |u^{(k)}|^2\,dt + \alpha^* \|u^{(n)}\|^2_{(-\infty,-x)} + C^* \|u\|^2_{(-\infty,-}$$

$$0 < \gamma < 1\,, \quad 0 < \alpha^* < \alpha, \quad u \in C_0^\infty(-\infty, \infty)\,.$$

hold.

3) There are constants b_k, $0 \leqslant k \leqslant n$, $b_n > 0$, such that the integrals

$$\int_{-\infty}^{\infty} [a_k(x) - b_k]\, dx, \quad 0 \leqslant k \leqslant n-1\,,$$

converge to numbers γ_k, and the integrals

$$\gamma_n = \int_{-\infty}^{\infty} [a_n(x) - b_n]\, dx, \quad \gamma_k = \int_{-\infty}^{\infty} a_k(x)\, dx, \quad n+1 \leqslant k \leqslant n^*\,,$$

are absolutely convergent.

4) As $|\omega| \to 0$

$$\sup_{-\infty < x < \infty} \int_{x}^{x+|\omega|} |a_k(t)|\, dt = o(1), \quad 0 \leqslant k \leqslant n-1\,,$$

and, if $n^* = 2n$,

$$\lim_{|x| \to \infty} a_{2n}(x) = 0\,.$$

Then A_0 is semibounded from below and all self-adjoint extensions A of A_0 have the same essential spectrum

$$\sigma_e(A) = [\Lambda, \infty), \quad \Lambda = \inf_{-\infty < t < \infty} \sum_{k=0}^{n} b_k t^{2k}\,.$$

If

$$\sum_{k=0}^{n^*} \gamma_k t_0^{2k} < 0 \quad \text{with } \Lambda = \sum_{k=0}^{n} b_k t_0^{2k}\,,$$

then any self-adjoint extension A has at least one eigenvalue below Λ.

Proof

By the hypothesis of the theorem all the conditions in Note 2.17 are satisfied, so that the essential spectrum is $[\Lambda, \infty)$ as asserted in the theorem. So it remains to show that the spectrum of any self-adjoint extension of A_0 is not empty below Λ. For this purpose $(A_0 u, u)$ is estimated for suitable u. We use the function $\eta(t)$ which arises from

$$x(t) = \begin{cases} 1, & |t| \leqslant \frac{3}{2}\,, \\ 0, & |t| > \frac{3}{2}\,, \end{cases}$$

by smoothing with the radius $\frac{1}{2}$ and we define

$$u_R(x) = \eta(x/R)e^{it_0 x} \in C_0^\infty(-\infty, \infty) .$$

By the addition of a suitable constant to $a_0(x)$ and b_0 we can make

$$\Lambda = \sum_{k=0}^{n} b_k t_0^{2k} = 0 ,$$

and this is assumed to be the case in what follows. We have

$$\begin{aligned}(A_0 u_R, u_R) &= \sum_{k=0}^{n} \int_{-\infty}^{\infty} b_k \,|u_R^{(k)}|^2 \,\mathrm{d}x \\ &+ \sum_{k=0}^{n} \int_{-\infty}^{\infty} [a_k(x) - b_k]\,|u_R^{(k)}|^2 \,\mathrm{d}x + \sum_{k=n+1}^{n} \int_{-\infty}^{\infty} a_k(x)|u_R^{(k)}|^2 \,\mathrm{d}x \qquad (80)\\ &= \sum_{k=0}^{n} b_k t_0^{2k} \int_{-\infty}^{\infty} \eta^2(x/R)\,\mathrm{d}x + \omega_1(R) + \sum_{k=0}^{n} t_0^{2k} \int_{-\infty}^{\infty} [a_k(x) - b_k]\eta^2(x/R)\,\mathrm{d}x \\ &+ \omega_2(R) + \sum_{k=n+1}^{n^*} t_0^{2k} \int_{-\infty}^{\infty} a_k(x)\eta^2(x/R)\,\mathrm{d}x + \omega_3(R) ,\end{aligned}$$

where the terms $\omega_i(R)$, $i = 1, 2, 3$, contain the derivatives of $\eta(x/R)$ which arise in each case. Clearly, as $R \to \infty$,

$$\int_{-\infty}^{\infty} [a_n(x) - b_n]\,\eta^2(x/R)\,\mathrm{d}x \to \gamma_n$$

and

$$\int_{-\infty}^{\infty} a_k(x)\eta^2(x/R)\,\mathrm{d}x \to \gamma_k, \quad n+1 \leqslant k \leqslant n^* .$$

The integrals

$$\int_{-\infty}^{\infty} [a_k(x) - b_k]\,\eta^2(x/R)\,\mathrm{d}x, \quad 0 \leqslant k \leqslant n-1 .$$

are dealt with by using a mean value theorem of integral calculus.† Since $\eta^2(x/R)$ is increasing on $(-\infty, -R)$, there exists a point ξ_1, $-2R \leqslant \xi_1 \leqslant -R$, by the mean value theorem, such that

† If $f(x)$ is integrable on $[a, b]$ and $\varphi(x)$ is monotone there, then there exists a point $\xi \in [a, b]$ such that $\int_a^b f(x)\varphi(x)\mathrm{d}x = \varphi(a) \int_a^\xi f(x)\mathrm{d}x + \varphi(b) \int_\xi^b f(x)\mathrm{d}x$.

$$\int_{-\infty}^{-R} [a_k(x) - b_k]\eta^2(x/R)\,dx = \int_{\xi_1}^{-R} [a_k(x) - b_k]\,dx, \quad \xi_1 = \xi_1(k, R).$$

Hence

$$\int_{-\infty}^{0} [a_k(x) - b_k]\eta^2(x/R)\,dx = \int_{\xi_1}^{0} [a_k(x) - b_k]\,dx,$$

and likewise we find a point $\xi_2, R \leqslant \xi_2 \leqslant 2R$, such that

$$\int_{0}^{\infty} [a_k(x) - b_k]\eta^2(x/R)\,dx = \int_{0}^{\xi_2} [a_k(x) - b_k]\,dx, \quad \xi_2 = \xi_2(k, R).$$

Hence

$$\sum_{k=0}^{n} t_0^{2k} \int_{-\infty}^{\infty} [a_k(x) - b_k]\eta^2(x/R)\,dx \to \sum_{k=0}^{n} \gamma_k t_0^{2k}$$

as $R \to \infty$. Now (80) simplifies to

$$(A_0 u_R, u_R) = \sum_{k=0}^{n^*} \gamma_k t_0^{2k} + o(1) + \sum_{i=1}^{3} \omega_i(R).$$

We show further that $\omega_i(R) = o(1)$ as $R \to \infty$, $i = 1, 2, 3$. In

$$\omega_1(R) = \sum_{0<\rho+\sigma\leqslant 2n} \int_{-\infty}^{\infty} C_{\rho\sigma} R^{-(\rho+\sigma)} \eta^{(\rho)}(x/R)\eta^{(\sigma)}(x/R)\,dx$$

the $C_{\rho\sigma}$ are constants and the derivatives are with respect to the variable $t = R^{-1}x$. We have

$$R^{-(\rho+\sigma)} \int_{-\infty}^{\infty} \eta^{(\rho)}(x/R)\eta^{(\sigma)}(x/R)\,dx = C_{\rho\sigma} R^{-(\rho+\sigma-1)},$$

where

$$\int_{-\infty}^{\infty} \eta^{(\rho)}(t)\eta^{(\sigma)}(t)\,dt = 0,$$

when $\rho + \sigma = 1$, and hence $\omega_1(R) = o(1)$ as $R \to \infty$. With regard to $\omega_2(R)$, integrals of the form

$$R^{-(\rho+\sigma)} \int_{-\infty}^{\infty} [a_k(x) - b_k]\eta^{(\rho)}(x/R)\eta^{(\sigma)}(x/R)\,dx, \quad \rho, \sigma \leqslant k \leqslant n, \quad \rho + \sigma > 0,$$

must be estimated. The support of the integrand lies in $[-2R, -R] \cup [R, 2R]$. We deal with the interval $[R, 2R]$ and split it into subintervals

$$[\xi_\nu, \xi_{\nu+1}], \quad \xi_0 = R < \xi_1 < \cdots < \xi_N = 2R,$$

on which the factor $\eta^{(\rho)}(x/R)\,\eta^{(\sigma)}(x/R)$ is monotone in each case. If the values $\xi'_\nu \in [\xi_\nu, \xi_{\nu+1}]$ are chosen suitably, by the mean value theorem we obtain

$$\int_R^{2R} [a_k(x) - b_k]\,\eta^{(\rho)}(x/R)\eta^{(\sigma)}(x/R)\,\mathrm{d}x$$

$$= \sum_{\nu=0}^{N-1} \int_{\xi_\nu}^{\xi_{\nu+1}} [a_k(x) - b_k]\,\eta^{(\rho)}(x/R)\eta^{(\sigma)}(x/R)\,\mathrm{d}x$$

$$= \sum_{\nu=0}^{N-1} \left\{ \eta^{(\rho)}(\xi_\nu/R)\eta^{(\sigma)}(\xi_\nu/R) \int_{\xi_\nu}^{\xi'_\nu} [a_k(x) - b_k]\,\mathrm{d}x \right.$$

$$\left. + \eta^{(\rho)}(\xi_{\nu+1}/R)\eta^{(\sigma)}(\xi_{\nu+1}/R) \int_{\xi'_\nu}^{\xi_{\nu+1}} [a_k(x) - b_k]\,\mathrm{d}x \right\}.$$

Because of the convergence of the integral

$$\int_{-\infty}^{\infty} [a_k(x) - b_k]\,\mathrm{d}x,$$

$$\int_{\xi_\nu}^{\xi'_\nu} [a_k(x) - b_k]\,\mathrm{d}x \to 0 \quad \text{and} \quad \int_{\xi'_\nu}^{\xi_{\nu+1}} [a_k(x) - b_k]\,\mathrm{d}x \to 0$$

as $R \to \infty$, and so

$$\int_R^{2R} [a_k(x) - b_k]\,\eta^{(\rho)}(x/R)\eta^{(\sigma)}(x/R)\,\mathrm{d}x = o(1).$$

The interval $[-2R, -R]$ can be dealt with similarly, and hence $\omega_2(R) = o(1)$ as $R \to \infty$. It is clear from the absolute convergence of the integrals

$$\int_{-\infty}^{\infty} a_k(x)\,\mathrm{d}x, \quad n+1 \leqslant k \leqslant n^*,$$

that $\omega_3(R) = o(1)$ as $R \to \infty$. Since

$$(A_0 u_R, u_R) = \sum_{k=0}^{n^*} \gamma_k t_0^{2k} + o(1)$$

as $R \to \infty$, we have $(A_0 u_R, u_R) < 0$ for sufficiently large R. Hence there must be at least one negative point of the spectrum of any self-adjoint extension A of A_0. Since $\Lambda = 0$ such points must be isolated eigenvalues. The restriction $\Lambda = 0$ can now be omitted, and Theorem 2.18 is proved.

In the case where $n = 1, a_1(x) = 1$ and $[0, \infty)$ is the basic interval, Putnam [47] has proved that the essential spectrum is $[0, \infty)$ if $a_0(x)$ is continuous and

$$\int_0^\infty a_0(t)\,\mathrm{d}t$$

converges.

Theorem 2.19

Let the conditions of Theorem 2.16 be satisfied, so that by this theorem the essential spectrum of any self-adjoint extension A of A_0 is the interval

$$[\Lambda, \infty), \quad \Lambda = \inf_{0<t<\infty} \sum_{k=0}^{n} b_k t^{2k} .$$

If in addition

$$\lim_{\xi \to \infty} \sum_{k=0}^{n^*} t_0^{2k} \int_0^\xi [a_k(t) - b_k]\,\mathrm{d}t = -\infty , \tag{81}$$

where t_0 is determined by

$$\Lambda = \sum_{k=0}^{n} b_k t_0^{2k} ,$$

and $b_k = 0$, $n + 1 \leqslant k \leqslant n^*$, then below Λ there is a countable set of eigenvalues of A which accumulate at Λ.

Proof

As in the proof of Theorem 2.18 we assume that $\Lambda = 0$ without loss of generality. We use the function

$$\eta(x) = \eta(\xi_1, \xi_2; x), \quad 1 < \xi_1, 1 + \xi_1 < \xi_2 < \infty ,$$

which arises from

$$\chi(\xi_1, \xi_2; x) = \begin{cases} 1, \xi_1 \leqslant x \leqslant \xi_2 , \\ \\ 0, \text{otherwise} , \end{cases}$$

by smoothing with radius $\frac{1}{2}$. Then with

$$u(x) = u(\xi_1, \xi_2; x) = \eta(\xi_1, \xi_2; x)e^{it_0 x}, \quad 0 \leqslant x < \infty,$$

we have

$$(A_0 u, u) = \sum_{k=0}^{n} b_k \int_0^\infty |u^{(k)}|^2 \,dx$$

$$+ \sum_{k=0}^{n} \int_0^\infty [a_k(x) - b_k] |u^{(k)}|^2 \,dx + \sum_{k=n+1}^{n^*} \int_0^\infty a_k(x) |u^{(k)}|^2 \,dx$$

$$= \sum_{k=0}^{n} b_k t_0^{2k} \int_0^\infty \eta^2(x) \,dx + \Omega_1 + \sum_{k=0}^{n} t_0^{2k} \int_0^\infty [a_k(x) - b_k] \eta^2(x) \,dx$$

$$+ \Omega_2(\xi_1, \xi_2) + \sum_{k=n+1}^{n^*} t_0^{2k} \int_0^\infty a_k(x) \eta^2(x) \,dx + \Omega_3(\xi_1, \xi_2),$$

where Ω_1 does not depend on ξ_1 and ξ_2, and $\Omega_2(\xi_1, \xi_2)$ is a sum of terms

$$c(t_0, \sigma, \tau) \int_0^\infty [a_k(x) - b_k] \eta^{(\sigma)}(x) \eta^{(\tau)}(x) \,dx$$

$$= c(t_0, \sigma, \tau) \left\{ \int_{|x-\xi_1|<1/2} [a_k(x) - b_k] \eta^{(\sigma)}(x) \eta^{(\tau)}(x) \,dx \right.$$

$$\left. + \int_{|x-\xi_2|<1/2} [a_k(x) - b_k] \eta^{(\sigma)}(x) \eta^{(\tau)}(x) \,dx \right\}, \quad \sigma + \tau > 0.$$

Since, by hypothesis

$$\left| \int_x^{x+1} [a_k(t) - b_k] \,dt \right| \leqslant C, \quad x \in [0, \infty), \quad 0 \leqslant k \leqslant n,$$

we have $|\Omega_2(\xi_1, \xi_2)| \leqslant C$ and similarly for Ω_3, where C is independent of ξ_1 and ξ_2. Hence

$$(A_0 u, u) = \sum_{k=0}^{n^*} t_0^{2k} \int_0^\infty [a_k(x) - b_k] \eta^2 \,dx + \Omega_4(\xi_1, \xi_2), \quad |\Omega_4| \leqslant C.$$

Then, by the mean value theorem as in the proof of Theorem 2.18, we obtain

$$(A_0 u, u) = \sum_{k=0}^{n^*} t_0^{2k} \int_{\xi_{2,k}}^{\xi_{2,k}} [a_k(x) - b_k]\, dx + \Omega_4(\xi_1, \xi_2)$$

$$= \sum_{k=0}^{n^*} t_0^{2k} \int_{\xi_1}^{\xi_2} [a_k(x) - b_k]\, dx + \Omega_5(\xi_1, \xi_2),$$

$$\xi_i - \tfrac{1}{2} \leqslant \xi_{ik} \leqslant \xi_i + \tfrac{1}{2}, \quad i = 1, 2, \quad |\Omega_5(\xi_1, \xi_2)| \leqslant C.$$

From (81) it now follows that $(A_0 u, u) < 0$ if ξ_1 is fixed and ξ_2 is chosen sufficiently large . Then we put $\xi_3 = \xi_2 + 1$ and choose ξ_4 so large that $(A_0 u, u) < 0$ similarly for $u = u(\xi_3, \xi_4; x)$. This method is continued and yields a countable set of functions

$$u = u(\xi_{2\nu-1}, \xi_{2\nu}; x) \in D(A_0), \quad \nu = 1, 2, \ldots,$$

for which $(A_0 u, u) < 0$ in each case. Hence below $\Lambda = 0$ there are an infinity of points of the spectrum of any self-adjoint extension A (see Theorem 5.2), and Theorem 2.19 is proved.

THE EULER DIFFERENTIAL OPERATOR

Another differential operator with a known spectrum is the Euler differential operator. It is generated by the differential expression

$$l[u] = \sum_{k=0}^{n} (-1)^k a_k (x^{2k} u^{(k)})^{(k)}, \quad 0 < x < \infty, \quad a_k = \text{const}, \quad a_n > 0.$$

We prove the following

Theorem 2.20
The operator

$$L_0 u = l[u], \quad D(L_0) = C_0^\infty(0, \infty),$$

is semibounded from below and essentially self-adjoint. The essential spectrum of the closure $\bar{L}_0$ of L_0 is the interval

$$[\Lambda, \infty), \quad \Lambda = \inf_{0<\xi<\infty} \sum_{k=0}^{n} a_k \prod_{j=1}^{k} [\xi^2 + (j - \tfrac{1}{2})^2]. \tag{82}$$

Proof
The transformation

$$(\tau u)(t) = e^{(1/2)t}u(e^t) \equiv \psi(t), \quad (\tau^{-1}\psi)(x) = u(x), \quad x = e^t,$$

defines an unitary operator τ, which transforms $L_2(0,\infty)$ onto $L_2(-\infty,\infty)$ and $C_0^\infty(0,\infty)$ onto $C_0^\infty(-\infty,\infty)$. We determine the behaviour of L_0 under the transformation τ. We have

$$u'(x) = x^{-3/2}\psi'(t) - \tfrac{1}{2}x^{-3/2}\psi(t) = x^{-3/2}\left(\frac{d}{dt} - \tfrac{1}{2}\right)\psi(t), \quad u(x) \in C_0^\infty(0,\infty),$$

and by repeating the differentiation we obtain

$$u^{(k)}(x) = x^{-k-1/2}\left(\frac{d}{dt} - k + \tfrac{1}{2}\right)\left(\frac{d}{dt} - k + \tfrac{3}{2}\right) \cdots \left(\frac{d}{dt} - \tfrac{1}{2}\right)\psi(t). \quad (83)$$

From (83) it follows that

$$(x^{2k}u^{(k)})' = x^{k-3/2}\left(\frac{d^2}{dt^2} - (k - \tfrac{1}{2})^2\right)\left(\frac{d}{dt} - k + \tfrac{3}{2}\right) \cdots \left(\frac{d}{dt} - \tfrac{1}{2}\right)\psi(t), \quad 1 \leqslant k \leqslant n,$$

and by repeating the differentiation we have

$$(-1)^k a_k (x^{2k}u^{(k)})^{(k)} = a_k x^{-1/2}\left(-\frac{d^2}{dt^2} + (k - \tfrac{1}{2})^2\right) \cdots \left(-\frac{d^2}{dt^2} + (\tfrac{1}{2})^2\right)\psi(t).$$

Consequently for the differential expression $l[u]$ we obtain

$$\sum_{k=0}^{n} (-1)^k a_k (x^{2k}u^{(k)})^{(k)} = e^{-(1/2)t} \sum_{k=0}^{n} a_k \prod_{j=0}^{k}\left(-\frac{d^2}{dt^2} + (j - \tfrac{1}{2})^2\right)\psi(t).$$

Hence, defining

$$\mathscr{L}_0\psi = \sum_{k=0}^{n} a_k \prod_{j=1}^{k}\left(-\frac{d^2}{dt^2} + (j - \tfrac{1}{2})^2\right)\psi, \quad D(\mathscr{L}_0) = C_0^\infty(-\infty,\infty),$$

we have $L_0 = \tau^{-1}\mathscr{L}_0\tau$. Since the operator $\mathscr{L}_0$ is semibounded from below and essentially self-adjoint by Theorem 2.4, the same is also true of L_0 and, since the essential spectrum of $\mathscr{L}_0$ is the interval (82) by Theorem 2.5, this is also true of the essential spectrum of $\bar{L}_0$. This proves Theorem 2.20.

Note 2.21
If we decompose the operator L_0 at any point γ, $0<\gamma<\infty$, then we obtain the two operators

$$L_{(\gamma,\infty),0}u = l[u], \quad D(L_{(\gamma,\infty),0}) = C_0^\infty(\gamma, \infty),$$

and

$$L_{(0,\gamma),0}u = l[u], \quad D(L_{(0,\gamma),0}) = C_0^\infty(0, \gamma) .$$

Correspondingly $\mathscr{L}_0$ is decomposed at $\gamma^* = \log \gamma$:

$$\mathscr{L}_{(\gamma^*,\infty),0}, \quad D(\mathscr{L}_{(\gamma^*,\infty),0}) = C_0^\infty(\gamma^*, \infty),$$

$$\mathscr{L}_{(-\infty,\gamma^*),0}, \quad D(\mathscr{L}_{(-\infty,\gamma^*),0}) = C_0^\infty(-\infty, \gamma^*) .$$

Self adjoint extensions $\mathscr{L}_{(\gamma^*,\infty)}$, $\mathscr{L}_{(-\infty,\gamma^*)}$ of $\mathscr{L}_{(\gamma^*,\infty),0}$ and $\mathscr{L}_{(-\infty,\gamma^*),0}$ have the same essential spectrum $[\Lambda, \infty)$ with Λ as in (82). Finally the self-adjoint extensions

$$L_{(\gamma,\infty)} = \tau^{-1}\mathscr{L}_{(\gamma^*,\infty)}\tau, \quad L_{(0,\gamma)} = \tau^{-1}\mathscr{L}_{(-\infty,\gamma^*)}\tau ,$$

of $L_{(\gamma,\infty),0}$ and $L_{(0,\gamma),0}$ also have the essential spectrum $[\Lambda, \infty)$.

In what follows we apply Theorem 2.8 to the Euler differential operator to obtain perturbations which do not change the essential spectrum. The following theorem corresponds to Theorem 2.10.

Theorem 2.22
Let the (real-valued) coefficients of the operator

$$L_0 u = \sum_{k=0}^{n^*} (-1)^k (a_k(x)x^{2k}u^{(k)})^{(k)} ,$$

$$D(L_0) = C_0^\infty(a, \infty), \quad a>0, \quad n \leqslant n^* \leqslant 2n ,$$

satisfy the following conditions.

1) $a_k(x) \in W_2^k(a, X)$ for all $X>a$, $0 \leqslant k \leqslant n^*$, $\inf\limits_{a<x<\infty} a_n(x) = \alpha > 0$,

$a_{n^*}(x) > 0, a<x<\infty$.

2) For $x \leqslant a$ there is an estimate

$$\left| \sum_{k=n+1}^{n^*} \int_x^\infty a_k^-(t) t^{2k} \, |u^{(k)}|^2 \, dt \right|$$

$$\leqslant \gamma \sum_{k=n+1}^{n^*} \int_x^\infty a_k^+(t) t^{2k} \, |u^{(k)}|^2 \, dt + \alpha^* \int_x^\infty t^{2n} \, |u^{(n)}|^2 \, dt + C^* \int_x^\infty |u|^2 \, dt, \tag{84}$$

where

$$0 < \gamma < 1, 0 < \alpha^* < \alpha, C^* > 0,$$

$$a_k^-(x) = \min(a_k(x), 0), \quad a_k^+(x) = \max(a_k(x), 0), u \in C_0^\infty(a, \infty) .$$

3) There are real numbers b_k, $0 \leqslant k \leqslant n$, $b_n > 0$, such that

$$\lim_{x \to \infty} x^{-1} \int_x^{2x} | a_k(t) - b_k | \, dt = 0, \quad 0 \leqslant k \leqslant n ,$$

$$\lim_{x \to \infty} x^{-1} \int_x^{2x} | a_k(t) | \, dt = 0, \quad n+1 \leqslant k \leqslant n^* . \tag{85}$$

Also, if $n^* = 2n$, then

$$a_{2n}(x) \to 0, x \to \infty .$$

Then the essential spectrum of any self-adjoint extension of L_0 is the interval

$$[\Lambda, \infty), \quad \Lambda = \inf_{0 < t < \infty} \sum_{k=0}^{n} b_k \prod_{j=1}^{k} [t^2 + (j - \tfrac{1}{2})^2] .$$

Proof

The proof is similar to that of Theorem 2.10. Without loss of generality we suppose that $a = 1$. Let $\varphi(x)$ be the function which arises from

$$f(x) = \begin{cases} 1, x \geqslant 3X, \quad X > 1 , \\ 0, x < 3X, \end{cases}$$

by smoothing with radius X, and let $\psi(x)$ be the function which arises from

$$g(x) = \begin{cases} 1, x \geqslant \tfrac{3}{2}X , \\ 0, x < \tfrac{3}{2}X , \end{cases}$$

by smoothing with radius $\frac{1}{2}X$. Then the coefficients $a_k(x)$ are changed as follows:

$$a_{k,X}(x) = \varphi(x)a_k(x) + (1 - \varphi(x))b_k\,, \quad 1 \leqslant x \leqslant \infty\,, \quad 0 \leqslant k \leqslant n\,,$$

$$a^+_{k,X}(x) = \psi(x)a^+_k(x)\,, \quad 1 \leqslant x \leqslant \infty\,, \quad n+1 \leqslant k \leqslant n^*,$$

$$a^-_{k,X}(x) = \varphi(x)a^-_k(x)\,, \quad 1 \leqslant x < \infty\,, \quad n+1 \leqslant k \leqslant n^*\,,$$

and we define

$$L_{X,0}u = \sum_{k=0}^{n^*} (-1)^k (a_{k,X}(x)x^{2k}u^{(k)})^{(k)}\,, \quad D(L_{X,0}) = C_0^\infty(1, \infty)\,.$$

Then (84) is also satisfied by $L_{X,0}$ in place of L_0, and hence we have

$$\begin{aligned}
&(L_{X,0}u, u)\\
&\geqslant \sum_{k=n+1}^{n^*} \int_1^\infty (a^-_{k,X} + a^+_{k,X})x^{2k} |u^{(k)}|^2\,\mathrm{d}x + \tfrac{1}{2}(\alpha - \alpha^*)\alpha^{-1} \int_1^\infty a_{n,X}x^{2n} |u^{(n)}|^2\,\mathrm{d}x\\
&+ \tfrac{1}{2}(\alpha + \alpha^*)\int_1^\infty x^{2n} |u^{(n)}|^2\,\mathrm{d}x + \sum_{k=0}^{n-1} \int_1^\infty a_{k,X}x^{2k} |u^{(k)}|^2\,\mathrm{d}x\\
&\geqslant (1-\gamma) \sum_{k=n+1}^{n^*} \int_1^\infty a^+_{k,X}x^{2k} |u^{(k)}|^2\,\mathrm{d}x + \tfrac{1}{2}(\alpha - \alpha^*)\alpha^{-1} \int_1^\infty a_{n,X}x^{2n} |u^{(n)}|^2\,\mathrm{d}x\\
&+ \tfrac{1}{2}(\alpha - \alpha^*)\int_1^\infty x^{2n} |u^{(n)}|^2\,\mathrm{d}x + \sum_{k=0}^{n-1} \int_1^\infty a_{k,X}x^{2k} |u^{(k)}|^2\,\mathrm{d}x - C^* \int_1^\infty |u|^2\,\mathrm{d}x\,.
\end{aligned} \tag{86}$$

We now estimate a term

$$\int_1^\infty a_{k,X}(x)x^{2k} |u^{(k)}|^2\,\mathrm{d}x\,, \quad 0 \leqslant k \leqslant n-1\,,$$

and we use (85), (69), and transformations $x = 2^\nu t$, $\nu = 0, 1, \ldots$. We obtain

$$\left| \int_1^\infty a_{k,X}(x)x^{2k} |u^{(k)}|^2\,\mathrm{d}x \right| \leqslant \sum_{\nu=0}^{\infty} \int_{2^\nu}^{2^{\nu+1}} a_{k,X}(x)x^{2k} |u^{(k)}|^2\,\mathrm{d}x$$

$$\leqslant \sum_{\nu=0}^{\infty} \int_1^2 |a_{k,X}(2^\nu t)|\, t^{2k} \left|\frac{\mathrm{d}^k}{\mathrm{d}t^k} u(2^\nu t)\right|^2 2^\nu \,\mathrm{d}t$$

$$\leqslant \sum_{\nu=0}^{\infty} 2^{2n+\nu} \left(\max_{1\leqslant t\leqslant 2} \left|\frac{\mathrm{d}^k}{\mathrm{d}t^k} u(2^\nu t)\right|^2\right) \int_1^2 |a_{k,X}(2^\nu t)|\,\mathrm{d}t$$

$$\leqslant 2^{2n} \sum_{\nu=0}^{\infty} 2^\nu \left(\epsilon \int_1^2 \left|\frac{\mathrm{d}^n}{\mathrm{d}t^n} u(2^\nu t)\right|^2 \mathrm{d}t + C_\epsilon \int_1^2 |u(2^\nu t)|^2 \,\mathrm{d}t\right) \int_1^2 |a_{k,X}(2^\nu t)|\,\mathrm{d}t$$

$$\leqslant 2^{2n} \sum_{\nu=0}^{\infty} \left(\epsilon \int_{2^\nu}^{2^{\nu+1}} 2^{2n\nu} |u^{(n)}(x)|^2 \,\mathrm{d}x + C_\epsilon \int_{2^\nu}^{2^{\nu+1}} |u(x)|^2 \,\mathrm{d}x\right) \int_{2^\nu}^{2^{\nu+1}} |a_{k,X}(x)|\, 2^{-\nu} \,\mathrm{d}x$$

$$\leqslant C \sum_{\nu=0}^{\infty} \left(\epsilon \int_{2^\nu}^{2^{\nu+1}} x^{2n} |u^{(n)}|^2 \,\mathrm{d}x + C_\epsilon \int_{2^\nu}^{2^{\nu+1}} |u(x)|^2 \,\mathrm{d}x\right)$$

$$= C\epsilon \int_1^\infty x^{2n} |u^{(n)}(x)|^2 \,\mathrm{d}x + CC_\epsilon \int_1^\infty |u(x)|^2 \,\mathrm{d}x\,. \tag{87}$$

If (87) is used in (86) and at the same time ϵ is chosen sufficiently small, then we obtain

$$(L_{X,0}u, u) \geqslant (1-\gamma) \sum_{k=n+1}^{n^*} \int_1^\infty (a_{k,X}^+(x) x^{2k} |u^{(k)}|^2 \,\mathrm{d}x$$
$$+ \tfrac{1}{2}(\alpha-\alpha^*)\alpha^{-1} \int_1^\infty a_{n,X}(x) x^{2n} |u^{(n)}|^2 \,\mathrm{d}x$$
$$+ \tfrac{1}{3}(\alpha-\alpha^*) \int_1^\infty x^{2n} |u^{(n)}|^2 \,\mathrm{d}x - C\int_1^\infty |u|^2 \,\mathrm{d}x\,. \tag{88}$$

We compare $L_{X,0}$ with the operator

$$B_0 u = \sum_{k=0}^{n} (-1)^k b_k (x^{2k} u^{(k)})^{(k)}, \quad D(B_0) = C_0^\infty(1,\infty)\,.$$

From (A.18) and (88) we obtain

$$(B_0 u, u) \leqslant C_1 \left(\int_1^\infty x^{2n} |u^{(n)}|^2 \,\mathrm{d}x + \int_1^\infty |u|^2 \,\mathrm{d}x\right) \leqslant C_2 (L_{X,0}u, u) + C_3 \|u\|^2\,.^\dagger$$

Therefore a sufficiently large constant can be added to $a_0(x)$ and b_0 that, for the operators B_0 and $L_{X,0}$ so altered, we have

$$\|u\|^2 \leqslant (B_0 u, u) \leqslant C(L_{X,0} u, u), \quad u \in C_0^\infty(1,\infty) \tag{89}$$

for every $X > a$. We assume that (89) holds in what follows.

† In the present proof the symbols $\|\cdot\|$, $(\cdot,\cdot)$ mean $\|\cdot\|_{(1,\infty)}$ and $(\cdot,\cdot)_{(1,\infty)}$ respectively.

To prove that $D(\hat{B}) \subseteq H_{L_{X,0}}$, let $\hat{u}(x)$ be any function in

$$D(\hat{B}) = D(B_0^*) \cap H_{B_0},$$

where

$$D(B_0^*) = \{u \mid u(x) \in W_2^{2n}(1,\xi) \text{ for all } \xi > 1 \text{ and } \int_1^\infty (x^{4n} \mid u^{(2n)} \mid^2 + \mid u \mid^2)\,\mathrm{d}x < \infty\}$$

and

$$H_{B_0} = \{u \mid u(x) \in W_2^n(1,\xi) \text{ for all } \xi > 1, \int_1^\infty (x^{2n} \mid u^{(n)} \mid^2 + \mid u \mid^2)\,\mathrm{d}x < \infty \text{ and } u^{(\nu)}(1) = 0,\ 0 \leqslant \nu \leqslant n-1\}.$$

The function

$$\hat{u}_1(x) = \varphi(x)\hat{u}(x), \quad 1 \leqslant x < \infty,$$

can be approximated in $H_{L_{X,0}}$ by C_0^∞-functions since

$$(L_{X,0}u, u) \leqslant C \int_1^\infty (x^{4n} \mid u^{(2n)} \mid^2 + \mid u \mid^2)\mathrm{d}x, \quad u \in H_{L_{X,0}}.$$

Hence $\hat{u}_1(x) \in H_{L_{X,0}}$. The function

$$\hat{u}_2(x) = (1 - \varphi(x))\hat{u}(x), \quad 1 \leqslant x < \infty,$$

clearly has the properties

$$\hat{u}_2(x) \in W_2^{2n}(1,\infty), \quad \hat{u}_2^{(\nu)}(1) = 0, \quad 0 \leqslant \nu \leqslant n-1,$$

$$\hat{u}_2(x) = 0, \qquad x > 4X,$$

and therefore can also be approximated in $H_{L_{X,0}}$ by C_0^∞-functions since

$$a_{k,X}(x) = 0, \quad 1 \leqslant x \leqslant X, \quad n+1 \leqslant k \leqslant n^*.$$

Then the sum $\hat{u}(x) = \hat{u}_1(x) + \hat{u}_2(x)$ also belongs to $H_{L_{X,0}}$.

The coefficients of the form

$$s_X[u,v] = \sum_{k=0}^{n^*} \int_1^\infty \eta_{k,X}(x)u^{(k)}\bar{v}^{(k)}\mathrm{d}x, \quad u, v \in H_{L_{X,0}},$$

are defined by

$$\eta_{0,X}(x) = \mid a_{0,X}(x) - b_0 \mid + (X+x)^{-1},$$

$\eta_{k,X}(x) = |a_{k,X}(x) - b_k|, \quad 1 \leqslant k \leqslant n, \quad \eta_{k,X}(x) = |a_{k,X}(x)|, \quad n+1 \leqslant k \leqslant n^*.$

Then we have

$$
\begin{aligned}
s_X[u,u] &= \sum_{k=0}^{n^*} \int_1^\infty \eta_{k,X}(x) x^{2k} \,|\, u^{(k)} \,|^2 \,\mathrm{d}x \\
&\leqslant \sum_{k=n+1}^{n^*} \int_1^\infty (a_{k,X}^+(x) - a_{k,X}^-(x)) x^{2k} \,|\, u^{(k)} \,|^2 \,\mathrm{d}x \qquad (90)\\
&+ \int_1^\infty a_{n,X}(x) x^{2n} |\, u^{(n)} \,|^2 \,\mathrm{d}x + C \int_1^\infty x^{2n} \,|\, u^{(n)} \,|^2 \,\mathrm{d}x + C \int_1^\infty |\, u \,|^2 \,\mathrm{d}x \\
&\leqslant (1+\gamma) \sum_{k=n+1}^{n^*} \int_1^\infty a_{k,X}^+(x) x^{2k} \,|\, u^{(k)} \,|^2 \,\mathrm{d}x + \int_1^\infty a_{n,X}(x) x^{2n} \,|\, u^{(n)} \,|^2 \,\mathrm{d}x \\
&\qquad + \widetilde{C} \int_1^\infty (x^{2n} \,|\, u^{(n)} \,|^2 + |\, u \,|^2) \mathrm{d}x
\end{aligned}
$$

which by (88) gives

$$
s_X[u,u] \leqslant \max\left(\frac{1+\gamma}{1-\gamma}, \frac{2\alpha}{\alpha - \alpha^*}, \frac{3\widetilde{C}}{\alpha - \alpha^*}\right) \cdot (L_{0,X} u, u) + C \,\|\, u \,\|^2 .
$$

Since $L_{X,0}$ is positive definite, we have

$$
s_X[u,u] \leqslant C \,\|\, u \,\|^2_{L_{X,0}}, \quad u \in H_{L_{X,0}} . \qquad (91)
$$

If, further, we put

$$
S_0 = L_{X,0} - B_0, \quad D(S_0) = C_0^\infty(1,\infty),
$$

then the above working shows that (15)–(18) of Theorem 2.6 are satisfied with $A_0 = L_{X,0}$. Therefore, as in the proof of Theorem 2.10, the operator T_X can be defined for which we have

$$
T_X \hat{B}^{-1} = \hat{B}^{-1} - \hat{L}_X^{-1} .
$$

To apply Theorem 2.8 we show now that

$$
\|\, \hat{B}^{-1} - \hat{L}_X^{-1} \,\| \to 0, \quad X \to \infty .
$$

Let h be any element of $L_2(1, \infty)$ with $\| h \| = 1$. If we define $u = \hat{B}^{-1} h$ and choose C_0 sufficiently large, then by (A.19) we obtain

$$1 = \| \hat{B} u \| = \| \sum_{k=0}^{n} (-1)^k b_k (x^{2k} u^{(k)})^{(k)} \|$$

$$\geqslant \| b_n x^{2n} u^{(2n)} \| - C_0 \sum_{j=0}^{2n-1} \| x^j u^{(j)} \|$$

$$\geqslant b_n \| x^{2n} u^{(2n)} \| - \epsilon \| x^{2n} u^{(2n)} \| - C_\epsilon \| u \| .$$

Taking $\epsilon = \frac{1}{2} b_n$, we get

$$1 \geqslant \tfrac{1}{2} b_n \| x^{2n} u^{(2n)} \| - C \| u \|$$

and so

$$\| x^{2n} u^{(2n)} \| \leqslant C(1 + \| u \|) .$$

Since

$$\| u \| = \| \hat{B}^{-1} h \| \leqslant \| \hat{B}^{-1} \|$$

we obtain

$$\int_1^\infty (x^{4n} | u^{(2n)} |^2 + | u |^2) \, dx \leqslant C, \quad u \in \hat{B}^{-1} h, \| h \| = 1 \tag{92}$$

Further, by (26) with $T_X, L_{X,0}$ and $s_X[u, u]$ in place of T, A_0 and $s[u, u]$ respectively and with $u = \hat{B}^{-1} h$, we obtain

$$\| (\hat{B}^{-1} - \hat{L}_X^{-1}) h \|^2 = \| T_X \hat{B}^{-1} h \|^2 = \| T_X u \|^2$$

$$\leqslant C_1 \| T_X u \|^2_{\hat{L}_{X,0}} \leqslant C_2 s_X[u, u] = C_2 \sum_{k=0}^{n^*} \int_1^\infty \eta_{k,X}(x) x^{2k} | u^{(k)} |^2 \, dx \tag{93}$$

$$\leqslant C_2 X^{-1} \| u \|^2 + C_2 \sum_{k=0}^{n^*} \int_X^\infty \eta_{k,X}(x) x^{2k} | u^{(k)} |^2 \, dx .$$

We now introduce the intervals

$$(X, \nu) = \{ x \mid 2^\nu X \leqslant x \leqslant 2^{\nu+1} X \} , \quad \nu = 0, 1, 2, \ldots ,$$

and, using (A.19), we estimate

$$\sum_{k=0}^{n^*} \int_X^\infty \eta_{k,X}(x) x^{2k} \mid u^{(k)} \mid^2 \mathrm{d}x$$

as follows:

$$\sum_{k=0}^{n^*} \int_X^\infty \eta_{k,X}(x) x^{2k} \mid u^{(k)} \mid^2 \mathrm{d}x = \sum_{k=0}^{n^*} \sum_{\nu=0}^{\infty} \int_{(X,\nu)} \eta_{k,X}(x) x^{2k} \mid u^{(k)} \mid^2 \mathrm{d}x$$

$$\leqslant \sum_{k=0}^{2n-1} \sum_{\nu=0}^{\infty} \int_1^2 \eta_{k,X}(2^\nu X t) t^{2k} \left| \left(\frac{\mathrm{d}}{\mathrm{d}t}\right)^k u(2^\nu X t) \right|^2 2^\nu X \, \mathrm{d}t$$

$$+ \sum_{\nu=0}^{\infty} \left(\max_{x \in (X,\nu)} \eta_{2n,X}(x) \right) \int_{(X,\nu)} x^{4n} \mid u^{(2n)}(x) \mid^2 \mathrm{d}x \;\dagger$$

$$\leqslant CX \sum_{k=0}^{2n-1} \sum_{\nu=0}^{\infty} 2^\nu \left(\int_1^2 \eta_{k,X}(2^\nu X t)\, \mathrm{d}t \right) \left(\int_1^2 \left(\left| \left(\frac{\mathrm{d}}{\mathrm{d}t}\right)^{2n} u(2^\nu X t) \right|^2 + \mid u(2^\nu X t)^2 \mid \right) \mathrm{d}t \right)$$

$$+ \sum_{\nu=0}^{\infty} \left(\max_{x \in (X,\nu)} \eta_{2n,X}(x) \right) \int_{(X,\nu)} x^{4n} \mid u^{(2n)} \mid^2 \mathrm{d}x$$

$$\leqslant C \sum_{k=0}^{2n-1} \sum_{\nu=0}^{\infty} (2^\nu X)^{-1} \left(\int_{(X,\nu)} \eta_{k,X}(x)\, \mathrm{d}x \right) \left(\int_{(X,\nu)} (x^{4n} \mid u^{(2n)} \mid^2 + \mid u \mid^2)\, \mathrm{d}x \right)$$

$$+ \sum_{\nu=0}^{\infty} \left(\max_{x \in (X,\nu)} \eta_{2n,X}(x) \right) \int_{(X,\nu)} x^{4n} \mid u^{(2n)} \mid^2 \mathrm{d}x \,. \tag{94}$$

If $\epsilon (> 0)$ is arbitrary, then by (85) X can be chosen so large that

$$(2^\nu X)^{-1} \left(\int_{(X,\nu)} \eta_{k,X}(x)\, \mathrm{d}x \right) < \epsilon, \quad \max_{x \in (X,\nu)} \eta_{2n,\, X}(x) < \epsilon\,, \tag{95}$$

$$\nu = 0, 1, \ldots, \quad 0 \leqslant k \leqslant 2n-1\,.$$

Then (94) simplifies to

$$\sum_{k=0}^{n^*} \int_X^\infty \eta_{k,X}(x) x^{2k} \mid u^{(k)}(x) \mid^2 \mathrm{d}x$$

$$\leqslant 2nC\epsilon \int_X^\infty (x^{4n} \mid u^{(2n)} \mid^2 + \mid u \mid^2)\, \mathrm{d}x + \epsilon \int_X^\infty x^{4n} \mid u^{(2n)} \mid^2 \mathrm{d}x, \quad X > X_0(\epsilon)\,.$$

† In the case $n \leqslant n^* < 2n$ we have to write $\eta_{k,X}(x) = 0$, $n^* + 1 \leqslant k \leqslant 2n$.

Thus, by (92), we obtain

$$\sum_{k=0}^{n^*} \int_X^\infty \eta_{k,X}(x) x^{2k} \mid u^{(k)} \mid^2 \mathrm{d}x \leqslant \epsilon\, C, \quad X > X_0(\epsilon), \quad u = \hat{B}^{-1} h . \quad (96)$$

If (96) is now used in (93), then it follows that, for a given positive integer ν, a point X_ν exists such that

$$\| \hat{B}^{-1} - \hat{L}_{X_\nu}^{-1} \| \leqslant \nu^{-1} .$$

With the decomposition method, which we have already used frequently, we can easily prove that

$$\sigma_e(\hat{L}) = \sigma_e(\hat{L}_X), \quad X > 1 .$$

Finally, since $\sigma_e(\hat{B})$ coincides with the interval given in the theorem (see Note 2.21), the theorem follows from Theorem 2.8.

Note 2.23
Condition (84) is satisfied if for instance there is a $C_0 > 0$, such that

$$x^{-1} \int_x^{2x} \mid a_k^-(t) \mid \mathrm{d}t \leqslant C_0 \cdot \min_{t \in [x, 2x]} a_{n^*}(t), \quad x \in [a, \infty), \quad a > 0 , \quad (97)$$

$$n + 1 \leqslant k < n^* .$$

Proof
If we take

$$(X, \nu) = \{x \mid 2^\nu X \leqslant x \leqslant 2^{\nu+1} X\} , \quad \nu = 0, 1 . \ldots ,$$

as in the proof of Theorem 2.22 and estimate as in (87), then by (97) we obtain

$$\int_{(X,\nu)} \mid a_k^-(x) \mid x^{2k} \mid u^{(k)} \mid^2 \mathrm{d}x \quad (98)$$

$$\leqslant \epsilon \int_{(X,\nu)} a_{n^*}(x) x^{2n^*} \mid u^{(n^*)} \mid^2 \mathrm{d}x + C_\epsilon \int_{(X,\nu)} \mid u \mid^2 \mathrm{d}x,$$

$$n + 1 \leqslant k \leqslant n^* - 1 .$$

Hence

$$\left| \sum_{k=n+1}^{n^*} \int_X^\infty a_k^-(x) x^{2k} \, | u^{(k)} |^2 \, dx \right|$$

$$\leqslant \epsilon n \int_X^\infty a_{n^*}(x) x^{2n^*} \, | u^{(n^*)} |^2 \, dx + n \, C_\epsilon \int_X^\infty | u |^2 \, dx \, .$$

Then (84) is satisfied if we choose $\epsilon < 1/n$.

Note 2.24
Similarly we can take $(0, a)$, $a > 0$, as the basic interval for the operator L_0. Then Theorem 2.22 remains valid if the hypotheses are changed to read as follows:

1) $a_k(x) \in W_2^k(X, a)$ for every $X, 0 < X < a, \quad 0 \leqslant k \leqslant n^*$,

$$a_{n^*}(x) > 0, \quad 0 < x < a, \quad n \leqslant n^* \leqslant 2n, \quad \inf_{0<x<a} a_n(x) = \alpha > 0 \, .$$

2) For all $0 < x \leqslant a$ there is an estimate

$$\left| \sum_{k=n+1}^{n^*} \int_0^x a_k^-(t) t^{2k} \, | u^{(k)} |^2 \, dt \right|$$

$$\leqslant \gamma \sum_{k=n+1}^{n^*} \int_0^x a_k^+(t) t^{2k} \, | u^{(k)} |^2 \, dt + \alpha^* \int_0^x t^{2n} \, | u^{(n)} |^2 \, dt + C^* \int_0^x | u |^2 \, dt \, ,$$

where

$$0 < \gamma < 1, \quad 0 < \alpha^* < \alpha, \quad C^* > 0, \quad u \in C_0^\infty(0, a) \, .$$

3) There are real numbers b_k, $\quad 0 \leqslant k \leqslant n, \quad b_n > 0$, such that

$$\lim_{x \to 0} x^{-1} \int_x^{2x} | a_k(t) - b_k | \, dt = 0, \quad 0 \leqslant k \leqslant n \, ,$$

$$\lim_{x \to 0} x^{-1} \int_x^{2x} | a_k(t) | \, dt = 0, \quad n + 1 \leqslant k \leqslant n^* \, .$$

Also, if $n^* = 2n$, then $a_{2n}(x) \to 0$ as $x \to 0$.

When we consider the operator L_0 with the basic interval $(0, \infty)$ the statement concerning the essential spectrum remains true if the conditions of Theorem 2.22 as well as those in Note 2.24 are satisfied.

Example 2.25

$$L_0 u = \sum_{k=0}^{2n} (-1)^k a_k (x^{\alpha_k} u^{(k)})^{(k)}, \quad a_k = \text{const}, \; u \in C_0^\infty(a, \infty) ;$$

$$\alpha_k = 2k, \quad -\infty < a_k < \infty, \quad 0 \leqslant k < n; \quad \alpha_n = 2n, \quad a_n > 0 ;$$

$$\alpha_k < 2k, \quad -\infty < a_k < \infty, \quad n+1 \leqslant k < 2n; \quad \alpha_{2n} < 4n, \quad a_{2n} > 0 ;$$

$$\alpha_{2n} \geqslant \max_{\substack{n+1 \leqslant k \leqslant 2n-1 \\ a_k < 0}} [\alpha_k + 2(2n-k)] \; . \tag{99}$$

Then we have

$$\sigma_e(L) = [\Lambda, \infty), \quad \Lambda = \inf_{0<t<\infty} \sum_{k=0}^{n} a_k \prod_{j=1}^{k} [t^2 + (j - \tfrac{1}{2})^2] \, ,$$

for every self-adjoint extension L of L_0.

Proof
L_0 is a perturbation of

$$B_0 u = \sum_{k=0}^{n} (-1)^k a_k (x^{2k} u^{(k)})^{(k)}, \quad u \in C_0^\infty(a, \infty) .$$

We show that the conditions of Theorem 2.22 are satisfied. First (97) follows from (99) since

$$x^{-1} \int_x^{2x} |a_k^-(t)| \, dt = x^{-1} \int_x^{2x} |a_k| \, t^{\alpha_k - 2k} \, dt \leqslant |a_k| \, x^{\alpha_k - 2k}$$

$$\leqslant C_1 x^{\alpha_{2n} - 4n} \leqslant C_2 \min_{t \in [x, 2x]} a_{2n} t^{\alpha_{2n} - 4n}, \quad n+1 \leqslant k < 2n, \quad a_k < 0 .$$

Then (84) follows from (97) as we have seen in Note 2.23. Also, (85) holds since

$$\lim_{x \to \infty} x^{-1} \int_x^{2x} |a_k(t)| \, dt = \lim_{x \to \infty} x^{-1} \int_x^{2x} |a_k| \, t^{\alpha_k - 2k} \, dt = 0, \quad n+1 \leqslant k < 2n \, ,$$

$$\lim_{x \to \infty} a_{2n}(x) = \lim_{x \to \infty} a_{2n} x^{\alpha_{2n} - 4n} = 0 \, .$$

If there is no negative $a_k, n+1 \leqslant k < 2n$, (99) is omitted.

If $(0, a), a > 0$, is the basic interval of L_0, then the conditions corresponding to (99) under which the same result concerning $\sigma_e(L)$ holds are:

$$\alpha_k = 2k, \quad -\infty < a_k < \infty, \quad 0 \leqslant k < n; \quad \alpha_n = 2n, \quad a_n > 0 \;;$$

$$\alpha_k > 2k, \quad -\infty < a_k < \infty, \quad n+1 \leqslant k < 2n; \quad \alpha_{2n} > 4n, \quad a_{2n} > 0 \;;$$

$$\alpha_{2n} \leqslant \min_{\substack{n+1 \leqslant k \leqslant 2n-1 \\ a_k < 0}} [\alpha_k + 2(2n-k)] \;.$$

Chapter 3

Discrete Spectra

The spectrum of an operator A is discrete if and only if $\sigma_e(A)$ is empty. In this chapter we obtain conditions on the coefficients of

$$a[u] = \sum_{k=0}^{n} (-1)^k (a_k(x) u^{(k)})^{(k)},$$

under which this is the case. The following results continue those of the previous chapter; especially we shall refer to Lemma 2.15. In this chapter we assume that the $a_k(x)$ satisfy the following conditions throughout.

i) $a_k(x) \in W_2^k(0, X)$, $0 \leqslant k \leqslant n$, for all $X > 0$.

ii) $a_n(x) > 0$, $0 \leqslant x \leqslant \infty$.

SUFFICIENT CONDITIONS

From Lemma 2.15 we obtain

Lemma 3.1

If $a_k(x) \geqslant 0$ on the interval $\omega = [x_1, x_2]$ for all k, then the inequalities

$$\sum_{k=r}^{s} \int_\omega a_k(x) |u^{(k)}|^2 \, dx \geqslant \left(\omega^{2(s-r)} \rho_{\omega,s}^{-1} + \sum_{k=r}^{s-1} |\omega|^{2(k-r)} \mu_{\omega,k}^{-1} \right)^{-1} \| u^{(r)} \|_\omega^2,$$

$$0 \leqslant r < s \leqslant n, \tag{1}$$

hold for all $u(x) \in C^n[x_1, x_2]$, where

$$\mu_{\omega,k} = \frac{1}{|\omega|} \int_\omega a_k(x) dx \quad \text{and} \quad \rho_{\omega,k} = \left(\frac{1}{|\omega|} \int_\omega \frac{dx}{a_k(x)} \right)^{-1}, \; 0 \leqslant k \leqslant n.$$

The estimates (1) are also true in the limiting cases $\mu_{\omega,k} = 0$ and $\rho_{\omega,s}^{-1} = \infty$; then the right-hand side of (1) is equal to zero in each case.

Proof
From (2.70) we have first

$$\sum_{k=r}^{s} \int_{\omega} a_k |u^{(k)}|^2 dx \geqslant \frac{1}{\mu_{\omega,s-1}^{-1} + |\omega|^2 \rho_{\omega,s}^{-1}} \|u^{(s-1)}\|_{\omega}^2 + \sum_{k=r}^{s-2} \int_{\omega} a_k |u^{(k)}|^2 dx.$$

Then we use (2.70) again, taking together the terms involving $u^{(s-1)}$ and $u^{(s-2)}$ on the right-hand side and, repeating this argument, we obtain the inequality (1).

Using Lemma 3.1 we prove

Theorem 3.2
Let $a_k(x) \leqslant 0$, $0 \leqslant k \leqslant n$. If

$$z = \{x_1 = 0, x_2, \ldots, x_\nu, \ldots\},\ \omega_\nu = [x_\nu, x_{\nu+1}],\ x_\nu < x_{\nu+1},$$

is a decomposition of the interval $[0, \infty)$ and we write

$$\Lambda_s = \sup_z \liminf_{\nu \to \infty} \left(|\omega_\nu|^{2s} \rho_{\omega_\nu,s}^{-1} + \sum_{k=0}^{s-1} |\omega_\nu|^{2k} \mu_{\omega_\nu,k}^{-1} \right)^{-1},\ 1 \leqslant s \leqslant n,$$

then, as regards the localisation of the essential spectrum of any self-adjoint extension A of A_0, where

$$A_0 u = a[u],\ D(A_0) = C_0^\infty(0, \infty),$$

we have

$$\sigma_e(A) \cap (-\infty, \Lambda_s) = \emptyset. \tag{2}$$

Proof
If any positive ϵ is given, we choose a decomposition z_ϵ such that

$$\left(|\omega_\nu|^{2s} \rho_{\omega_\nu,s}^{-1} + \sum_{k=0}^{s-1} |\omega_\nu|^{2k} \mu_{\omega_\nu,k}^{-1} \right)^{-1} > \Lambda_s - \epsilon \tag{3}$$

for $\nu \geqslant N(\epsilon)$. The operator A_0 is decomposed at the point $x_{N(\epsilon)}$. If as previously the Friedrichs extensions of the three operators A_0,

$$A_{(0,x_{N(\epsilon)}),0}\, u = a[u],\ D(A_{(0,x_{N(\epsilon)}),0}) = C_0^\infty(0, x_{N(\epsilon)}),$$

and

$$A_{(x_{N(\epsilon)},\infty),0}\, u = a[u],\ D(A_{(x_{N(\epsilon)},\infty),0}) = C_0^\infty(x_{N(\epsilon)}, \infty),$$

are denoted by $\hat{A}, \hat{A}_{(0,x_{N(\epsilon)})}$ and $\hat{A}_{(x_{N(\epsilon)},\infty)}$ respectively, we have

$$\sigma_e(\hat{A}) = \sigma_e(\hat{A}_{(0,x_{N(\epsilon)})}) \cup \sigma_e(\hat{A}_{(x_{N(\epsilon)},\infty)}).$$

Since there is an estimate of the form

$$\| u \|^2_{W_2^n(0,x_{N(\epsilon)})} \leqslant C \left(\sum_{k=0}^{n} \int_0^{x_{N(\epsilon)}} a_k(x) | u^{(k)} |^2 dx + \| u \|^2_{(0,x_{N(\epsilon)})} \right),$$

$$u \in H_{A(0,x_{N(\epsilon)}),0},$$

and the embedding of $W_2^n(0, x_{N(\epsilon)})$ into $L_2(0, x_{N(\epsilon)})$ is compact we obtain

$$\sigma_e(\hat{A}_{(0,x_{N(\epsilon)})}) = \emptyset,$$

and hence

$$\sigma_e(\hat{A}) = \sigma_e(\hat{A}_{(x_{N(\epsilon)},\infty)}). \tag{4}$$

On integrating by parts and then using the inequalities (1) and (3), we obtain

$$\begin{aligned}
&(\hat{A}_{(x_{N(\epsilon)},\infty)}u, u) \\
&\geqslant \sum_{k=0}^{s} \int_{x_{N(\epsilon)}}^{\infty} a_k(x) | u^{(k)} |^2 dx \geqslant \sum_{\nu=N(\epsilon)}^{\infty} \sum_{k=0}^{s} \int_{x_\nu}^{x_{\nu+1}} a_k(x) | u^{(k)} |^2 dx \\
&\geqslant \sum_{\nu=N(\epsilon)}^{\infty} \left(| \omega_\nu |^{2s} \rho^{-1}_{\omega_{\nu,s}} + \sum_{k=0}^{s-1} | \omega_\nu |^{2k} \mu^{-1}_{\omega_{\nu,k}} \right)^{-1} \| u \|^2_{\omega_\nu} \\
&\geqslant (\Lambda_s - \epsilon) \| u \|^2_{(x_{N(\epsilon)},\infty)}.
\end{aligned} \tag{5}$$

From (5) it follows that

$$\sigma_e(\hat{A}_{(x_{N(\epsilon)},\infty)}) \cap (-\infty, \Lambda_s - \epsilon) = \emptyset,$$

and therefore we obtain

$$\sigma_e(\hat{A}) \cap (-\infty, \Lambda_s - \epsilon) = \emptyset$$

by (4). The theorem follows since $\epsilon(>0)$ is arbitrary and the essential spectrum is the same for all self-adjoint extensions of A_0.

Corollary 3.3
If we have an integer s, $1 \leqslant s \leqslant n$, and a decomposition

$$z = \{x_1 = 0, x_2, \ldots, x_\nu, \ldots\}, \omega_\nu = [x_\nu, x_{\nu+1}],$$

of $[0, \infty)$ such that

$$|\omega_\nu|^{-2s} \rho_{\omega_\nu, s} \to \infty, \quad |\omega_\nu|^{-2k} \mu_{\omega_\nu, k} \to \infty, \quad 0 \leqslant k \leqslant s-1, \tag{6}$$

as $\nu \to \infty$, then the spectrum of any self-adjoint extension of A_0 is discrete.

Proof
From (6) we have $\Lambda_s = \infty$, and the corollary follows from Theorem 3.2.

Remark 3.4
The conditions of Corollary 3.3 are satisfied for instance when

$$a_s(x) \to \infty, \quad \int_x^{x+|\omega|} a_k(t)\mathrm{d}t \to \infty, \quad 0 \leqslant k \leqslant s-1,$$

as $x \to \infty$ for some s and a fixed $|\omega| > 0$.

In the next theorem we compare the operator A with the Euler differential operator and, for this, we refer to Theorem 2.20 and Note 2.21.

Theorem 3.5
Let there be integers r and s with $0 \leqslant r < s \leqslant n$, constants $\alpha_0, \ldots, \alpha_{r-1}$, and a point $X > 0$ such that

$$a_k(x) \geqslant 0, \quad r \leqslant k \leqslant n, \ a_k(x) \geqslant \alpha_k x^{2k}, \quad 0 \leqslant k \leqslant r-1, \quad x > X,$$

If

$$z = \{x_1 = 0, x_2, \ldots, x_\nu, \ldots\}, \quad \omega_\nu = [x_\nu, x_{\nu+1}], \quad x_\nu < x_{\nu+1},$$

is a decomposition of $[0, \infty)$ and we write

$$\Lambda_{r,s} = \sup_z \liminf_{\nu \to \infty} x_{\nu+1}^{-2r} \left(|\omega_\nu|^{2(s-r)} \rho_{\omega_\nu, s}^{-1} + \sum_{k=r}^{s-1} |\omega_\nu|^{2(k-r)} \mu_{\omega_\nu, k}^{-1}\right)^{-1} \tag{7}$$

where $\Lambda_{r,s} > 0$ is assumed, then we have

$$\sigma_e(A) \cap (-\infty, \beta) = \emptyset,$$

where

$$\beta = \inf_{0<\xi<\infty} \sum_{k=0}^{r} \alpha_k \prod_{j=1}^{s-1} [\xi^2 + (j-\tfrac{1}{2})^2]$$

and

$$\alpha_r = \Lambda_{r,s}.$$

In particular, if there is a decomposition z such that

$$x_{\nu+1}^{-2r} \mid \omega \mid^{2(r-s)} \rho_{\omega_{\nu},s} \to \infty \quad \text{and} \quad x_{\nu+1}^{-2r} \mid \omega_\nu \mid^{2(r-k)} \mu_{\omega_\nu,k} \to \infty, \qquad r \leqslant k \leqslant s-1, \tag{8}$$

as $\nu \to \infty$ then the spectrum of A is discrete. This is the case if, for instance,

$$x^{-2r} a_s(x) \to \infty, \quad x^{-2r} \int_x^{x+|\omega|} a_k(t)\mathrm{d}t \to \infty, \quad r \leqslant k \leqslant s-1, \tag{9}$$

as $x \to \infty$ for a fixed $\mid \omega \mid > 0$.
(In the case $r = 0$ we have Theorem 3.2.)

Proof
Let $\epsilon > 0$ be so small that $\Lambda_{r,s} - \epsilon > 0$. A decomposition z_ϵ is chosen such that

$$x_{\nu+1}^{-2r} (\mid \omega_\nu \mid^{2(s-r)} \rho_{\omega_\nu,s}^{-1} + \sum_{k=r}^{s-1} \mid \omega_\nu \mid^{2(k-r)} \mu_{\omega_\nu,k}^{-1})^{-1} > \Lambda_{r,s} - \epsilon \tag{10}$$

when $\nu \geqslant N(\epsilon)$, $x_{N(\epsilon)} > X$. As in the proof of Theorem 3.2 the operator is decomposed at the point $x_{N(\epsilon)}$. Using a corresponding notation, by (1) and (10), we obtain

$$(\hat{A}_{(x_{N(\epsilon)},\infty)} u, u) \geqslant \sum_{k=r}^{s} \int_{x_{N(\epsilon)}}^{\infty} a_k(x) \mid u^{(k)} \mid^2 \mathrm{d}x$$

$$+ \sum_{k=0}^{r-1} \alpha_k \int_{x_{N(\epsilon)}}^{\infty} x^{2k} \mid u^{(k)} \mid^2 \mathrm{d}x$$

$$= \sum_{\nu=N(\epsilon)}^{\infty} \sum_{k=r}^{s} \int_{x_\nu}^{x_{\nu+1}} a_k(x) \mid u^{(k)} \mid^2 \mathrm{d}x$$

$$+\sum_{k=0}^{r-1}\alpha_k\int_{x_{N(\epsilon)}}^{\infty}x^{2k}|u^{(k)}|^2\,dx$$

$$=\sum_{\nu=N(\epsilon)}^{\infty}\left(|\omega_\nu|^{2(s-r)}\rho_{\omega_\nu,s}^{-1}+\sum_{k=r}^{s-1}|\omega_\nu|^{2(k-r)}\mu_{\omega_\nu,k}^{-1}\right)^{-1}\|u^{(r)}\|_{\omega_\nu}^2$$

$$+\sum_{k=0}^{r-1}\alpha_k\int_{x_{N(\epsilon)}}^{\infty}x^{2k}|u^{(k)}|^2\,dx$$

$$\geqslant\sum_{\nu=N(\epsilon)}^{\infty}x_{\nu+1}^{2r}(\Lambda_{r,s}-\epsilon)\|u^{(r)}\|_{\omega_\nu}^2+\sum_{k=0}^{r-1}\alpha_k\int_{x_{N(\epsilon)}}^{\infty}x^{2k}|u^{(k)}|^2\,dx$$

$$\geqslant\sum_{k=0}^{r}\alpha_{k,\epsilon}\int_{x_{N(\epsilon)}}^{\infty}x^{2k}|u^{(k)}|^2\,dx,$$

where

$$\alpha_{r,\epsilon}=\Lambda_{r,s}-\epsilon>0,\quad \alpha_{k,\epsilon}=\alpha_k,\quad 0\leqslant k\leqslant n-1.$$

If the operator

$$B_{(x_{N(\epsilon)},\infty)}u=\sum_{k=0}^{r}(-1)^k\alpha_{k,\epsilon}(x^{2k}u^{(k)})^{(k)},\quad u\in C_0^\infty(x_{N(\epsilon)},\infty),$$

is considered, then by the last inequality we have

$$(\hat{A}_{(x_{N(\infty)},\infty)}u,u)_{(x_{N(\epsilon)},\infty)}\geqslant(\hat{B}_{(x_{N(\epsilon)},\infty)}u,u)_{(x_{N(\epsilon)},\infty)}.$$

Now we apply the Courant variational principle (Theorem 1.6). Since by Note 2.21 the spectrum of $\hat{B}_{(x_{N(\epsilon)},\infty)}$ is the interval

$$[\beta_\epsilon,\infty),\qquad \beta_\epsilon=\inf_{0<\xi<\infty}\sum_{k=0}^{r}\alpha_{k,\epsilon}\sum_{j=1}^{k}[\xi^2+(j-\tfrac{1}{2})^2],$$

we have

$$\sigma_e(\hat{A}_{(x_{N(\epsilon)},\infty)})\cap(-\infty,\beta_\epsilon)=\emptyset.$$

From (4) it follows that

$$\sigma_e(\hat{A}) \cap (-\infty, \beta_\epsilon) = \emptyset,$$

and letting $\epsilon \to 0$ we obtain

$$\sigma_e(\hat{A}) \cap (-\infty, \beta) = \emptyset.$$

This proves the first part of the theorem.

The second part follows from the fact that $\alpha_r = \Lambda_{r,s} \to \infty$ implies that $\beta \to \infty$. From (9) we obtain (8) if the decomposition $z = \{x_1, \ldots, x_\nu, \ldots\}$, with $x_\nu = (\nu - 1)|\omega|$, is used. This completes the proof.

NECESSARY AND SUFFICIENT CONDITIONS

In what follows necessary and sufficient conditions for the discreteness of the spectra are given where the coefficients $a_k(x)$ are restricted in certain ways. First we consider the case of non-negative coefficients.

Theorem 3.6

If the first three conditions of

i) $$\sup_{x \in [0,\infty)} \int_x^{x+1} \frac{dt}{a_n(t)} < \infty,$$

ii) There exists a point $X > 0$ such that $a_k(x) \geqslant 0, x > X, 0 \leqslant k \leqslant n-1$.

iii) There exists a number $\delta > 0$ such that

$$\liminf_{x \to \infty} \frac{1}{|\omega|} \int_x^{x+|\omega|} a_k(t)\, dt \geqslant \delta, \quad 1 \leqslant k \leqslant n-1,$$

for each fixed $|\omega| > 0$.

iv) $$\sup_{x \in [0,\infty)} \int_x^{x+1} a_k(t)\, dt < \infty, \quad 1 \leqslant k \leqslant n,$$

are satisfied, then the spectrum of any self-adjoint extension of A_0 is discrete provided that, in addition,

v) $$\lim_{x \to \infty} \int_x^{x+|\omega|} a_0(t) dt = \infty \quad \text{for each fixed } |\omega| > 0.$$

Under the conditions (i)–(iv) the spectrum is discrete if and only if (v) is satisfied.

Proof

If (i)–(iii) and (v) hold, then, by (1) (for $r = 0$ and $s = n$), corresponding to any $K > 0$ the value $|\omega|$ can be chosen so small that

$$\sum_{k=0}^{n} \int_{x}^{x+|\omega|} a_k(t)|u^{(k)}|^2 \, dt \geqslant K \int_{x}^{x+|\omega|} |u|^2 \, dt$$

for all sufficiently large x, $x \geqslant x_K$. Consequently

$$(A_{(x_K,\infty),0}u, u)_{(x_K,\infty)} \geqslant K \| u \|^2_{(x_K,\infty)},$$

where

$$A_{(x_K,\infty),0}u = a[u], \quad D(A_{(x_K,\infty),0}) = C_0^\infty(x_K, \infty),$$

from which we have

$$\sigma_e(\hat{A}) \cap (-\infty, K) = \emptyset.$$

Since K can be arbitrarily large, we obtain $\sigma_e(\hat{A}) = \emptyset$.

Now let the conditions (ii) and (iv) be satisfied. Then we show that $\sigma_e(\hat{A})$ is not empty if (v) is not satisfied. We use Theorem 1.2. If (v) does not hold, then there are a value $|\omega^*| > 0$ and a sequence $\omega_\nu = [x_\nu, x_\nu + |\omega^*|]$, $\nu = 1, 2, \ldots$, of mutually disjoint intervals for which

$$\sup_{\nu=1,2,\ldots} \int_{\omega_\nu} a_0(t) dt < \infty. \tag{11}$$

We choose a function $u_1(x) \in C_0^\infty(-\infty, \infty)$ with $\| u_1 \| = 1$, the support of which is contained in the interval ω_1. Defining $u_\nu(x) = u_1(x - x_\nu + x_1)$, $\nu = 1, 2, \ldots$, we obtain functions for which $(u_\nu, u_{\nu'}) = \delta_{\nu\nu'}$. The set $\{u_\nu\}_{\nu=1,2,\ldots}$ is not precompact in the Hilbert space H but it is bounded in the energy space H_{A_0}, as we now show. First,

$$(\hat{A}u_\nu, u_\nu) \leqslant \sum_{k=0}^{n} (\max |u_1^{(k)}(x)|^2) \int_{\omega_\nu} a_k(x) dx \leqslant C \sum_{k=0}^{n} \int_{\omega_\nu} a_k(x) dx.$$

Therefore by (iv) and (11) there is a constant C_0 independent of ν such that

$$(\hat{A}u_\nu, u_\nu) + \| u_\nu \|^2 = \| u_\nu \|^2_{H_{A_0}} \leqslant C_0, \quad \nu = 1, 2, \ldots.$$

We can now apply Theorem 1.2, and Theorem 3.6 follows.

Next, negative coefficients $a_k(x)$, $0 \leq k \leq n-1$, will also be considered. At the same time, however, we assume that $a_n(x)$ is bounded from below by a positive constant. First we prove a theorem which resembles Theorem 2.16.

Theorem 3.7[36]
Let the following conditions be satisfied.

i) $$a_n(x) \geq b_n > 0, \quad 0 \leq x < \infty.$$

ii) If $a_k^-(x)$ denotes the negative part of $a_k(x)$, then

$$\sup_{x \in [0,\infty)} \int_x^{x+|\omega|} |a_k^-(t)|\, dt \to 0 \quad \text{as} \quad |\omega| \to 0, \quad 0 \leq k \leq n-1.$$

iii) For each fixed $|\omega| > 0$

$$\liminf_{x \to \infty} \frac{1}{|\omega|} \int_x^{x+|\omega|} a_k(t)\, dt \geq b_k, \quad 0 \leq k \leq n-1$$

with finite constants b_k.
Then A_0, $D(A_0) = C_0^\infty(0, \infty)$, is bounded from below and, for any self-adjoint extension A of A_0,

$$\sigma_e(A) \cap (-\infty, \mu) = \emptyset,$$

where

$$\mu = \inf_{0 < \xi < \infty} \sum_{k=0}^{n} b_k \xi^{2k}.$$

Proof
We obtain the semiboundedness as follows. We have

$$(A_0 u, u) \geq b_n \| u^{(n)} \|^2 + \sum_{k=0}^{n-1} \int_0^\infty a_k^-(x) | u^{(k)}|^2\, dx$$

$$\geq b_n \| u^{(n)} \|^2 - \sum_{\nu=0}^{\infty} \sum_{k=0}^{n-1} \int_\nu^{\nu+1} |a_k^-(t)|\, | u^{(k)}|^2\, dt$$

$$\geq \sum_{\nu=0}^{\infty} \left\{ b_n \| u^{(n)} \|^2_{(\nu,\nu+1)} - \sum_{k=0}^{n-1} \left(\max_{x \in [\nu,\nu+1]} | u^{(k)}|^2 \right) \int_\nu^{\nu+1} |a_k^-(t)|\, dt \right\}.$$

By (ii) there is a constant C independent of ν and k such that

$$\int_{\nu}^{\nu+1} |a_k^-(t)|\,\mathrm{d}t \leqslant C.$$

Then, by (A.9) with a sufficiently small ϵ, we obtain

$$(A_0 u, u) \geqslant \sum_{\nu=0}^{\infty}\left(b_n \|u^{(n)}\|^2_{(\nu,\nu+1)} - C \sum_{k=0}^{n-1} \max_{x\in[\nu,\nu+1]} |u^{(k)}|^2 \right)$$

$$\geqslant \sum_{\nu=0}^{\infty} (\tfrac{1}{2} b_n \|u^{(n)}\|^2_{(\nu,\nu+1)} - C_{b_n} \|u\|^2_{(\nu,\nu+1)})$$

$$= \tfrac{1}{2} b_n \|u^{(n)}\|^2 - C_{b_n} \|u\|^2.$$

To prove that

$$\sigma_e(A) \cap (-\infty, \mu) = \emptyset$$

we choose any $\epsilon > 0$ with $0 < \epsilon < b_n$ and define the operators

$$B_{\epsilon,0} u = \sum_{k=0}^{n} (-1)^k (b_k - \epsilon) u^{(2k)}, \quad D(B_{\epsilon,0}) = C_0^\infty(0,\infty),$$

and

$$S_{\epsilon,0} u = \sum_{k=0}^{n} (-1)^k [(a_k(x) - b_k + \epsilon) u^{(k)}]^{(k)}, \quad D(S_{\epsilon,0}) = C_0^\infty(0,\infty).$$

Then we have

$$B_{\epsilon,0} + S_{\epsilon,0} = A_0\,.$$

Further, we decompose A_0, $B_{\epsilon,0}$, and $S_{\epsilon,0}$ at a point x_ϵ, $0 < x_\epsilon < \infty$, and define the operators

$$A_{(x_\epsilon,\infty),0} u = a[u], \quad D(A_{(x_\epsilon,\infty),0}) = C_0^\infty(x_\epsilon,\infty)\,,$$

$$B_{\epsilon,(x_\epsilon,\infty),0} u = \sum_{k=0}^{n} (-1)^k (b_k - \epsilon) u^{(2k)}, \quad D(B_{\epsilon,(x_\epsilon,\infty),0}) = C_0^\infty(x_\epsilon,\infty),$$

$$S_{\epsilon,(x_\epsilon,\infty),0} u = \sum_{k=0}^{n} (-1)^k [(a_k(x) - b_k + \epsilon) u^{(k)}]^{(k)}, \quad D(S_{\epsilon,(x_\epsilon,\infty),0}) = C_0^\infty(x_\epsilon,\infty)$$

Then we have

$$\sigma_e(\hat{A}_{(x_\epsilon,\infty)}) = \sigma_e(\hat{A}). \tag{12}$$

As regards $\hat{B}_{\epsilon,(x_\epsilon,\infty)}$, from the observations made in proving Theorem 2.5, it follows that

$$(\hat{B}_{\epsilon,(x_\epsilon,\infty)}u, u)_{(x_\epsilon,\infty)} \geqslant \mu_\epsilon \| u \|^2_{(x_\epsilon,\infty)}, \quad u \in D(\hat{B}_{\epsilon,(x_\epsilon,\infty)}), \tag{13}$$

where

$$\mu_\epsilon = \inf_{0<\xi<\infty} \sum_{k=0}^{n} (b_k - \epsilon)\xi^{2k}.$$

As regards $S_{\epsilon,(x_\epsilon,\infty),0}$, we show now that

$$(S_{\epsilon,(x_\epsilon,\infty),0}u, u)_{(x_\epsilon,\infty)} \geqslant 0, \quad u \in C_0^\infty(x_\epsilon, \infty), \tag{14}$$

when x_ϵ is chosen sufficiently large. For this, we introduce the notation

$$s_k(x) = a_k(x) - b_k + \epsilon, \quad 0 \leqslant k \leqslant n,$$

$$s_k^+(x) = \max(a_k(x) - b_k + \epsilon, 0),$$

$$s_k^-(x) = \min(a_k(x) - b_k + \epsilon, 0),$$

$$\sigma_{k,\nu}^+ = \frac{1}{|\omega|}\int_{x_\nu}^{x_\nu+|\omega|} s_k^+(t)\,dt, \quad \sigma_{k,\nu}^- = \frac{1}{|\omega|}\int_{x_\nu}^{x_\nu+|\omega|} s_k^-(t)\,dt,$$

$$\sigma_{k,\nu} = \frac{1}{|\omega|}\int_{x_\nu}^{x_\nu+|\omega|} s_k(t)\,dt, \quad x_{\nu+1} = x_\nu + |\omega|,$$

$$\nu = 1, 2, \ldots, \quad x_1 = x_\epsilon.$$

Using (2.73) and the relationship $\sigma_{n-1,\nu}^+ = \sigma_{n-1,\nu} - \sigma_{n-1,\nu}^-$, we obtain

$$(S_{\epsilon,(x_\epsilon,\infty),0}u, u)_{(x_\epsilon,\infty)} \geqslant \epsilon \| u^{(n)} \|^2_{(x_\epsilon,\infty)} + \sum_{k=0}^{n-1} \int_{x_\epsilon}^{\infty} s_k(x) |u^{(k)}|^2\,dx$$

$$= \sum_{\nu=1}^{\infty} \left(\epsilon \| u^{(n)} \|^2_{(x_\nu,x_{\nu+1})} + \int_{x_\nu}^{x_{\nu+1}} s_{n-1} |u^{(n-1)}|^2\,dx \right)$$

$$+ \sum_{k=0}^{n-2} \int_{x_\epsilon}^{\infty} s_k |u^{(k)}|^2\,dx$$

$$\geqslant \sum_{\nu=1}^{\infty} \frac{\sigma_{n-1,\nu}[1+4|\omega|^2\epsilon^{-1}\sigma_{n-1,\nu}^-]-4|\omega|^2\epsilon^{-1}(\sigma_{n-1,\nu}^-)^2}{[1+2|\omega|^2\epsilon^{-1}\sigma_{n-1,\nu}^-][1+2|\omega|^2\epsilon^{-1}\sigma_{n-1,\nu}-2|\omega|^2\epsilon^{-1}\sigma_{n-1,\nu}^-]} \times$$

$$\times \|u^{(n-1)}\|^2_{(x_\nu,x_{\nu+1})} + \sum_{k=0}^{n-2}\int_{x_\epsilon}^{\infty} s_k(x)|u^{(k)}|^2\,dx\,. \quad (15)$$

From (iii) it follows that

$$\liminf_{\nu\to\infty} \sigma_{n-1,\nu} \geqslant \epsilon\,.$$

As $|\omega|\to 0$, by (ii) all terms in (15) which contain the factor $\sigma_{n-1,\nu}^-$ tend to zero uniformly in ν. Hence

$$\frac{\sigma_{n-1,\nu}[1+4|\omega|^2\epsilon^{-1}\sigma_{n-1,\nu}^-]-4|\omega|^2\epsilon^{-1}(\sigma_{n-1,\nu}^-)^2}{[1+2|\omega|^2\epsilon^{-1}\sigma_{n-1,\nu}^-][1+2|\omega|^2\epsilon^{-1}\sigma_{n-1,\nu}-2|\omega|^2\epsilon^{-1}\sigma_{n-1,\nu}^-]} \geqslant \frac{\epsilon}{2},$$

$$0<\epsilon<b_n,\quad |\omega|<\delta(\epsilon),\quad \nu\geqslant N(\epsilon).$$

If now $x_\epsilon \geqslant x_{N(\epsilon)} = X_1$, then from (15) we have

$$(S_{\epsilon,(x_\epsilon,\infty),0}u,\, u)_{(x_\epsilon,\infty)} \geqslant \frac{\epsilon}{2}\|u^{(n-1)}\|^2_{(x_\epsilon,\infty)} + \sum_{k=0}^{n-2}\int_{x_\epsilon}^{\infty} s_k(x)|u^{(k)}|^2\,dx\,. \quad (16)$$

We now repeat the argument which led to (16) to obtain

$$(S_{\epsilon,(x_\epsilon,\infty),0}u,\, u)_{(x_\epsilon,\infty)} \geqslant \frac{\epsilon}{2^2}\|u^{(n-2)}\|^2_{(x_\epsilon,\infty)} + \sum_{k=0}^{n-3}\int_{x_\epsilon}^{\infty} s_k(x)|u^{(k)}|^2\,dx \quad (17)$$

with a new sufficiently large x_ϵ, $x_\epsilon \geqslant X_2 \geqslant X_1$.

Then after n steps we have the estimate

$$(S_{\epsilon,(x_\epsilon,\infty),0}u,\, u)_{(x_\epsilon,\infty)} \geqslant \frac{\epsilon}{2^n}\|u\|^2_{(x_\epsilon,\infty)},\quad x_\epsilon \geqslant X_n \geqslant \ldots \geqslant X_1. \quad (18)$$

Putting together (13) and (18) we obtain

$$(A_{(x_\epsilon,\infty),0}u,\, u)_{(x_\epsilon,\infty)} \geqslant \mu_\epsilon \|u\|^2_{(x_\epsilon,\infty)},\quad u\in C_0^\infty(x_\epsilon,\infty)\,, \quad (19)$$

and by (12) it follows that

$$\sigma_e(\hat{A}) \cap (-\infty, \mu_\epsilon) = \emptyset .$$

Since ϵ, $0 < \epsilon < b_n$, can be chosen arbitrarily, we have

$$\sigma_e(\hat{A}) \cap (-\infty, \mu) = \emptyset ,$$

and Theorem 3.7 is proved.

Theorem 3.7 leads to the next result.

Theorem 3.8

Let the following conditions be satisfied.

i) $$\inf_{0<x<\infty} a_n(x) = b_n > 0 .$$

ii) $$\sup_{0<x<\infty} \int_x^{x+1} |a_k(t)|\, dt < \infty, \quad 1 \leqslant k \leqslant n .$$

iii) $$\inf_{0<x<\infty} \int_x^{x+1} a_0^-(t)\, dt > -\infty .$$

Then the operator A_0 is bounded from below, and the spectrum of any self-adjoint extension of A_0 is discrete if and only if

iv) $$\lim_{x\to\infty} \int_x^{x+|\omega|} a_0(t)\, dt = \infty \tag{20}$$

for each fixed $|\omega| > 0$.

Proof

We start with

$$(A_0 u, u) = \int_0^\infty a_n(x) |u^{(n)}|^2\, dx + \sum_{k=0}^{n-1} \int_0^\infty [a_k^+(x) + a_k^-(x)] |u^{(k)}|^2\, dx .$$

From the proof of Theorem 3.7 we see that

$$\tfrac{1}{2} \int_0^\infty a_n(x) |u^{(n)}|^2\, dx + \sum_{k=0}^{n-1} \int_0^\infty a_k^-(x) |u^{(k)}|^2\, dx \geqslant -C_1 \|u\|^2$$

and hence

$$(A_0 u, u) \geqslant \tfrac{1}{2} \int_0^\infty a_n(x) |u^{(n)}|^2 \, dx + \sum_{k=0}^{n-1} \int_0^\infty a_k^+(x) |u^{(k)}|^2 \, dx - C_1 \|u\|^2. \tag{21}$$

From (iv) we have

$$\lim_{x \to \infty} \int_x^{x+|\omega|} a_0^+(t) \, dt = \infty$$

for each fixed $|\omega| > 0$. Hence, by (19), a point x_ϵ can be found for each $\tilde{b}_0 > 0$ such that

$$\tfrac{1}{2} \int_{x_\epsilon}^\infty a_n |u^{(n)}|^2 \, dx + \sum_{k=0}^{n-1} \int_{x_\epsilon}^\infty a_k^+ |u^{(k)}|^2 \, dx \geqslant \mu_\epsilon \|u\|^2_{(x_\epsilon, \infty)},$$

$$u \in C_0^\infty(x_\epsilon, \infty), \tag{22}$$

where

$$\mu_\epsilon = \inf_{0 < \xi < \infty} [(\tfrac{1}{2} b_n - \epsilon) \xi^{2n} - \epsilon \xi^{2n-2} - \cdots - \epsilon \xi^2 + \tilde{b}_0 - \epsilon],$$

$$0 < \epsilon < \tfrac{1}{2} b_n .$$

Then from (21) and (22) we obtain

$$\sigma_e(\hat{A}_{(x_\epsilon, \infty)}) \cap (-\infty, \mu_\epsilon - C_1) = \emptyset .$$

and hence

$$\sigma_e(\hat{A}) \cap (-\infty, \tilde{b}_0 - C_1) = \emptyset .$$

Since $\tilde{b}_0$ can be chosen arbitrarily large we obtain

$$\sigma_e(\hat{A}) = \emptyset .$$

If (20) is not satisfied, then the argument is as in the second part of the proof of Theorem 3.6. Theorem 3.8 is now proved.

In the special cases where

$$a_n(x) \equiv 1, \quad a_0^-(x) \equiv 0, \quad a_k(x) \equiv 0, \quad 1 \leqslant k \leqslant n-1,$$

Theorem 3.8 was first proved by Molčanov [35].

In the case $n = 1$ the criterion of Molčanov has been generalized by Brinck [6] as follows. In place of (iii) of Theorem 3.8 the weaker condition is imposed that there exists a finite C such that

$$\int_\Omega a_0(x)\, dx \geqslant -C$$

for all intervals $\Omega \subseteq [0, \infty)$ of size $|\Omega| \leqslant 1$. Then the spectrum of any self-adjoint extension of A_0,

$$A_0 u = -u'' + a_0(x)u, \quad D(A_0) = C_0^\infty(0, \infty),$$

is discrete if and only if (20) holds. A further weakening of the condition on $a_0(x)$ was given by Ismagilov [23] (see [15] § 59).

Next we consider differential operators whose coefficients $a_k(x)$ are restricted by powers of x, as in Theorem 3.5, where the differential operator was compared with the Euler differential operator.

Theorem 3.9

Let the following conditions be satisfied.

i) There are integers r and s, $0 \leqslant r < s \leqslant n$, a value α, $-\infty < \alpha \leqslant 2$, and a point $X > 0$ such that

$$a_k(x) \geqslant 0, \quad s \leqslant k \leqslant n, \quad a_s(x) \geqslant \sigma x^{\alpha s}, \quad \sigma > 0,$$

and

$$a_k(x) \geqslant c_k x^{\alpha k}, \quad -\infty < c_k < 0, \quad r \leqslant k < s,$$

for $x > X$.

ii) For each fixed $h > 0$,

$$x^{-(\tau+2r)} \int_x^{x+hx^\tau} a_r(t)\, dt \to \infty, \quad x \to \infty, \tag{23}$$

where $\tau = \frac{1}{2}\{\alpha - r(2-\alpha)\}$.

iii) The functions $a_k(x)x^{-2k}$, $x > X$, $0 \leqslant k < r$, are bounded from below by a constant γ.

Then the spectrum of any self-adjoint extension of A_0 is discrete.

Proof

We fix a value h, $0<h<\min(1,\frac{1}{2}\sigma)$, and choose $x=\xi_h$ so large that $\xi_h>\max(X,1)$ and

$$x^{-(\tau+2r)}\int_x^{x+hx^\tau} a_r(t)\,dt \geqslant 1, \quad \xi_h \leqslant x < \infty. \tag{24}$$

The operator A_0 is decomposed at $x=\xi_h$ and the quadratic form of

$$A_{(\xi_h,\infty),0}u=a[u], \quad D(A_{(\xi_h,\infty),0})=C_0^\infty(\xi_h,\infty),$$

is estimated from below. For this, we consider a decomposition of $[\xi_h,\infty)$ into intervals ω_ν which are defined inductively by

$$\omega_\nu=[x_\nu,x_{\nu+1}], \quad x_{\nu+1}=x_\nu+hx_\nu^\tau, \quad \nu=1,2,\ldots, \quad x_1=\xi_h. \tag{25}$$

Using (A.18), we have

$$(A_{(\xi_h,\infty),0}u,u)_{(\xi_h,\infty)}=\sum_{k=0}^{n}\int_{\xi_h}^\infty a_k(x)|u^{(k)}|^2\,dx$$

$$\geqslant \sigma\int_{\xi_h}^\infty x^{\alpha s}|u^{(s)}|^2\,dx+\sum_{k=r}^{s-1}c_k\int_{\xi_h}^\infty x^{\alpha k}|u^{(k)}|^2\,dx \pm \frac{\sigma}{2}\int_{\xi_h}^\infty x^{\alpha(r+1)}|u^{(r+1)}|^2\,dx$$

$$+\int_{\xi_h}^\infty a_r^+|u^{(r)}|^2\,dx+\gamma\sum_{k=0}^{r-1}\int_{\xi_h}^\infty x^{2k}|u^{(k)}|^2\,dx$$

$$\geqslant \frac{\sigma}{2}\int_{\xi_h}^\infty x^{\alpha(r+1)}|u^{(\tau+1)}|^2\,dx+\int_{\xi_h}^\infty a_r^+(x)|u^{(r)}|^2\,dx$$

$$+\gamma\sum_{k=0}^{r-1}\int_{\xi_h}^\infty x^{2k}|u^{(k)}|^2\,dx - C_\sigma\|u\|^2_{(\xi_h,\infty)}. \tag{26}$$

To estimate the first two terms on the right-hand side the inequality (1) and the decomposition (25) are used. We begin with the case $0\leqslant\alpha\leqslant 2$. We have

$$\frac{\sigma}{2}\int_{\xi_h}^\infty x^{\alpha(r+1)}|u^{(r+1)}|^2\,dx+\int_{\xi_h}^\infty a_r^+(x)|u^{(r)}|^2\,dx$$

$$\geqslant \sum_{\nu=1}^{\infty}(|\omega_\nu|^2 2\sigma^{-1}x_\nu^{-\alpha(r+1)}+\mu^{-1}_{\omega_\nu,r})^{-1}\|u^{(r)}\|^2_{\omega_\nu}$$

$$= \sum_{\nu=1}^{\infty} x_\nu^{2r} \left\{ h^2 2\sigma^{-1} + h x_\nu^{\tau+2r} \left(\int_{\omega_\nu} a_r(x)\, \mathrm{d}x \right)^{-1} \right\}^{-1} \| u^{(r)} \|^2_{\omega_\nu}$$

$$\geqslant \sum_{\nu=1}^{\infty} x_\nu^{2r} x_{\nu+1}^{-2r} \left\{ h^2 2\sigma^{-1} + h x_\nu^{\tau+2r} \left(\int_{\omega_\nu} a_r(x)\, \mathrm{d}x \right)^{-1} \right\}^{-1} \int_{\omega_\nu} x^{2r} \, | u^{(r)} |^2 \, \mathrm{d}x \, . \tag{27}$$

From (25) and the choice of h, it follows that

$$x_\nu^{2r} x_{\nu+1}^{-2r} \geqslant (\tfrac{1}{2})^{2r}, \quad \nu = 1, 2, \ldots .$$

Using this and (24) in (27), we obtain

$$\frac{\sigma}{2} \int_{\xi_h}^{\infty} x^{\alpha(r+1)} | u^{(r+1)} |^2 \, \mathrm{d}x + \int_{\xi_h}^{\infty} a_r^+ \, | u^{(r)} |^2 \, \mathrm{d}x \geqslant (\tfrac{1}{2})^{2r+1} h^{\ 1} \int_{\xi_h}^{\infty} x^{2r} | u^{(r)} |^2 \, \mathrm{d}x. \tag{28}$$

We now write $\gamma_r = (\frac{1}{2})^{2r+1} h^{-1}$ and $\gamma_k = \gamma$, $0 \leqslant k \leqslant r - 1$. Then from (26) and (28) we have

$$(A_{(\xi_h, \infty), 0} u, u)_{(\xi_h, \infty)} \geqslant \sum_{k=0}^{r} \gamma_k \int_{\xi_h}^{\infty} x^{2k} | u^{(k)} |^2 \, \mathrm{d}x - C_\sigma \| u \|^2_{(\xi_h, \infty)} \, . \tag{29}$$

As in the proof of Theorem 3.5, from (29) we obtain

$$\sigma_{\mathrm{e}} (\hat{A}_{(\xi_h, \infty)}) \cap (-\infty, \beta_h) = \varnothing \, , \tag{30}$$

where

$$\beta_h = \inf_{0 < \xi < \infty} \sum_{k=0}^{r} \gamma_k \prod_{j=1}^{r} [\xi^2 + (j - \tfrac{1}{2})^2] - C_\sigma \, ,$$

for the essential spectrum of the Friedrichs extension $\hat{A}_{(\xi_h, \infty)}$ of $A_{(\xi_h, \infty), 0}$. From (30) and the relationship

$$\sigma_{\mathrm{e}} (\hat{A}_{(\xi_h, \infty)}) = \sigma_{\mathrm{e}} (\hat{A}) \, ,$$

which we have used often, it follows that

$$\sigma_{\mathrm{e}} (\hat{A}) \cap (-\infty, \beta_h) = \varnothing \, .$$

Since γ_r tends to infinity as $h \to 0$, and thus β_h also becomes arbitrarily large, we have $\sigma_e(\hat{A}) = \emptyset$.

Now we consider the case $\alpha < 0$. There is similar working to that leading to (27), and we now have

$$\frac{\sigma}{2} \int_{\xi_h}^{\infty} x^{\alpha(r+1)} |u^{(r+1)}|^2 \, dx + \int_{\xi_h}^{\infty} a_r^+(x) |u^{(r)}|^2 \, dx$$

$$\geqslant \sum_{\nu=1}^{\infty} (|\omega_\nu|^2 2\sigma^{-1} (x_\nu + |\omega_\nu|)^{|\alpha|(r+1)} + \mu_{\omega_\nu,r}^{-1})^{-1} \| u^{(r)} \|_{\omega_\nu}^2$$

$$\geqslant \sum_{\nu=1}^{\infty} (|\omega_\nu|^2 \eta x_\nu^{-\alpha(r+1)} + \mu_{\omega_\nu,r}^{-1})^{-1} \| u^{(r)} \|_{\omega_\nu}^2 ,$$

where
$$\eta = \sigma^{-1} 2^{|\alpha|(r+1)+1} > 0.$$

Thus the connection with what was considered in the case $0 \leqslant \alpha \leqslant 2$ is established, and Theorem 3.9 follows in the case $\alpha < 0$ also.

We consider some special cases.

1) If $\alpha = 0$ (23) becomes

$$x^{-r} \int_x^{x+hx^{-r}} a_r(t) \, dt \to \infty, \quad x \to \infty, \tag{31}$$

for each fixed $h > 0$ and, in this case and under the conditions $a_k(x) \geqslant 0, r+1 \leqslant k \leqslant n-1$, Theorem 3.9 has been proved by Hinton and Lewis [20] with $s = n$.

2) In the case $\alpha = 2$ we have $\tau = 1$, and (23) becomes

$$x^{-(1+2r)} \int_x^{x+hx} a_r(t) \, dt \to \infty, \quad x \to \infty, \tag{32}$$

for each fixed $h > 0$. (32) is satisfied if for instance $a_r(x) x^{-2r} \to \infty$ as $x \to \infty$. Eastham [9] has pointed out the discreteness of the spectrum in this case.

3) If we take $\alpha = 2r/(r+1)$, then $\tau = 0$ and (23) becomes

$$x^{-2r} \int_x^{x+h} a_r(t) \, dt \to \infty, \quad x \to \infty. \tag{33}$$

Next we formulate necessary and sufficient conditions for the discreteness of the spectrum of differential operators whose coefficients $a_k(x)$ may increase as $x^{\alpha k}$ as $x \to \infty$. The next theorem complements Theorem 3.9.

Theorem 3.10
Let there exist an integer s, $1 \leqslant s \leqslant n$, a value α, $-\infty < \alpha \leqslant 2$, a point $X > 0$, and constants c_k, C_k with

$$0 \leqslant C_k < \infty, \quad s \leqslant k \leqslant n, \quad 0 < c_s < C_s < \infty,$$
$$-\infty < c_k < C_k < \infty, \quad 0 < k < s, \quad -\infty < c_0 ,$$

such that

$$0 \leqslant a_k(x) \leqslant C_k x^{\alpha k}, \quad s \leqslant k \leqslant n, \quad c_s x^{\alpha s} \leqslant a_s(x) \leqslant C_s x^{\alpha s},$$
$$c_k x^{\alpha k} \leqslant a_k(x) \leqslant C_k x^{\alpha k}, \quad 0 < k < s, \quad c_0 \leqslant a_0(x) ,$$

for $x > X$. Then the spectrum of any self-adjoint extension of A_0 is discrete if and only if

$$\lim_{x \to \infty} x^{-(\alpha/2)} \int_x^{x + h x^{\alpha/2}} a_0(t)\, dt = \infty \quad \text{for each fixed} \quad h > 0. \tag{34}$$

Proof
If we put $r = 0$ in (23) then $\tau = \alpha/2$, and we obtain (34). Consequently the spectrum of A is discrete by Theorem 3.9.

We now prove the necessity of (34) for the discreteness of the spectrum. If (34) is not satisfied then there are numbers $h^* > 0$ and $C^* < \infty$ such that for a countable number of mutually disjoint intervals

$$\omega_\nu = [\xi_\nu, \xi_\nu + h^* \xi_\nu^{\alpha/2}], \quad \nu = 1, 2, \ldots, \tag{35}$$

the inequality

$$\xi_\nu^{-(\alpha/2)} \int_{\xi_\nu}^{\xi_\nu + h^* \xi_\nu^{\alpha/2}} a_0(t)\, dt \leqslant C^* \tag{36}$$

is true. From (36), writing $\xi_\nu^* = \xi_\nu + h^* \xi_\nu^{\alpha/2}$, we have

$$\xi_\nu^{-(\alpha/2)} \int_{\xi_\nu}^{\xi_\nu^*} a_0^+(t)\, dt = \xi_\nu^{-(\alpha/2)} \int_{\xi_\nu}^{\xi_\nu^*} a_0(t)\, dt - \xi_\nu^{-(\alpha/2)} \int_{\xi_\nu}^{\xi_\nu^*} a_0^-(t)\, dt$$
$$\leqslant C^* + \xi_\nu^{-(\alpha/2)} \int_{\xi_\nu}^{\xi_\nu^*} | a_0^-(t) |\, dt \leqslant C^* + h^* | c_0 | . \tag{37}$$

We choose a real-valued function $\varphi(x)$, $-\infty < x < \infty$, with $\| \varphi \| = 1$, the support of which is contained in $(0, 1)$, and define

$$u_\nu(x) = h^{*-(1/2)} \xi_\nu^{-(\alpha/4)} \varphi[h^{*-1} \xi_\nu^{-(\alpha/2)} (x - \xi_\nu)], \quad \nu = 1, 2, \ldots.$$

Then we have

$$\| u_\nu \|^2 = h^{*-1} \xi_\nu^{-(\alpha/2)} \int_{\omega_\nu} \varphi^2 [h^{*-1} \xi_\nu^{-(\alpha/2)} (x - \xi_\nu)] \, dx = 1$$

and, for $k \geqslant 0$, it follows that

$$[u_\nu^{(k)}(x)]^2 \leqslant h^{*-2k-1} \xi_\nu^{-(\alpha/2)(2k+1)} \max_{0<t<1} [\varphi^{(k)}(t)]^2$$

$$\leqslant \frac{\Phi_k}{| \omega_\nu |^{2k+1}} \tag{38}$$

with

$$| \omega_\nu | = h^* \xi_\nu^{\alpha/2} . \tag{39}$$

Now A_0 is semibounded from below and, to prove the boundedness of the set $\{u_\nu\}_{\nu=1,2,\ldots}$ in H_{A_0}, it is sufficient to estimate $(A_0 u_\nu, u_\nu)$. We begin with the case $0 \leqslant a \leqslant 2$. By (37), (38), and (39) we obtain

$$(A_0 u_\nu, u_\nu) \leqslant \sum_{k=1}^{n} C_k \int_{\omega_\nu} x^{\alpha k} | u_\nu^{(k)} |^2 \, dx + \int_{\omega_\nu} a_0^+(x) | u_\nu |^2 \, dx$$

$$\leqslant \sum_{k=1}^{n} C_k \Phi_k | \omega_\nu |^{-(2k+1)} \int_{\omega_\nu} x^{\alpha k} \, dx + \Phi_0 | \omega_\nu |^{-1} \int_{\omega_\nu} a_0^+(x) \, dx$$

$$\leqslant \sum_{k=1}^{n} C_k \Phi_k | \omega_\nu |^{-2k} (\xi_\nu + h^* \xi_\nu^{\alpha/2})^{\alpha k} + \Phi_0 [h^{*-1} C^* + | c_0 |]$$

$$\leqslant C \sum_{k=1}^{n} | \omega_\nu |^{-2k} \xi_\nu^{\alpha k} + \Phi_0 [h^{*-1} C^* + | c_0 |]$$

$$\leqslant C \sum_{k=1}^{n} h^{*-2k} + \Phi_0 (h^{*-1} C^* + | c_0 |) ,$$

which is independent of ν. If $\alpha < 0$, then we use

$$\int_{\omega_\nu} x^{\alpha k} \, dx \leqslant | \omega_\nu | \xi_\nu^{\alpha k}$$

to obtain a similar result.

Since the H_{A_0}-bounded set $\{u_\nu\}_{\nu=1,2,\ldots}$ is not pre-compact in H because of $(u_\nu, u_{\nu'}) = \delta_{\nu,\nu'}$, $\nu,\nu' = 1, 2, \ldots$, the essential spectrum of $\hat{A}$ is not empty by Theorem 1.2. This is then true for all self-adjoint extensions of A_0, and Theorem 3.10 is proved.

In the case $\alpha = 2$ we give a further necessary and sufficient condition for the discreteness of the spectrum, which involves a middle coefficient $a_r(x)$, $1 \leqslant r \leqslant s-1$.

Theorem 3.11
Let there exist integers r and s, $0 < r < s \leqslant n$, a point $X > 0$, and constants c_k, C_k with

$$0 \leqslant C_k < \infty, \quad s \leqslant k \leqslant n, \quad 0 < c_s < C_s < \infty,$$
$$-\infty < c_k < C_k < \infty, \quad 0 \leqslant k < s, \quad k \neq r, \quad -\infty < c_r\,,$$

such that

$$0 \leqslant a_k(x) \leqslant C_k x^{2k}, \quad s \leqslant k \leqslant n, \quad c_s x^{2s} \leqslant a_s(x) \leqslant C_s x^{2s},$$
$$c_k x^{2k} \leqslant a_k(x) \leqslant C_k x^{2k}, \quad 0 \leqslant k < s, \quad k \neq r, \quad c_r x^{2r} \leqslant a_r(x)\,,$$

for $x > X$. Then the spectrum of any self-adjoint extension of A_0 is discrete if and only if

$$\lim_{x \to \infty} x^{-(2r+1)} \int_x^{x+hx} a_r(t)\,\mathrm{d}t = \infty \tag{40}$$

for each fixed $h > 0$.

Proof
In (23) we now have $\tau = 1$ and the discreteness of the spectrum follows from Theorem 3.9. If (40) is not satisfied, then there are numbers $h^* > 0$ and $C^* < \infty$ such that

$$\xi_\nu^{-(2r+1)} \int_{\xi_\nu}^{\xi_\nu + h^* \xi_\nu} a_r(t)\,\mathrm{d}t \leqslant C^*, \quad \nu = 1, 2, \ldots, \tag{41}$$

where we assume that

$$\xi_\nu + h^* \xi_\nu < \xi_{\nu+1}, \quad \nu = 1, 2, \ldots.$$

From (41) we have

$$\xi_\nu^{-(2r+1)} \int_{\xi_\nu}^{\xi_\nu+h^*\xi_\nu} a_r^+(t)\,dt = \xi_\nu^{-(2r+1)} \int_{\xi_\nu}^{\xi_\nu+h^*\xi_\nu} a_r(t)\,dt + \xi_\nu^{-(2r+1)} \int_{\xi_\nu}^{\xi_\nu+h^*\xi_\nu} |\,a_r^-(t)\,|\,dt$$

$$\leqslant C^* + |\,c_r\,|\xi_\nu^{-(2r+1)} \int_{\xi_\nu}^{\xi_\nu+h^*\xi_\nu} t^{2r}\,dt$$

$$\leqslant C^* + |\,c_r\,|\xi_\nu^{-2r} h^*(\xi_\nu + h^*\xi_\nu)^{2r} = C^* + |\,c_r\,| h^*(1+h^*)^{2r}\,. \tag{42}$$

Now, using the functions $u_\nu(x)$ defined in the proof of Theorem 3.10 for the special case $\alpha = 2$ and using (38) and (42), we obtain

$$(A_0 u_\nu, u_\nu) \leqslant \sum_{k=0,\ldots,n}^{k\neq r} C_k \int_{\omega_\nu} x^{2k}\,|\,u_\nu^{(k)}|^2\,dx + \int_{\omega_\nu} a_r^+(x)|\,u_\nu^{(r)}\,|^2\,dx$$

$$\leqslant \sum_{k=0,\ldots,n}^{k\neq r} \Phi_k C_k\,|\,\omega_\nu\,|^{-(2k+1)} \int_{\omega_\nu} x^{2k}\,dx + \Phi_r |\,\omega_\nu\,|^{-(2r+1)} \int_{\omega_\nu} a_r^+(x)\,dx$$

$$\leqslant \sum_{k=0,\ldots,n}^{k\neq r} \Phi_k C_k (h^*\xi_\nu)^{-2k}(\xi_\nu + h^*\xi_\nu)^{2k} + \Phi_r h^{*-(2r+1)}[C^* + |\,c_r\,| h^*(1+h^*)^{2r}]$$

$$= \sum_{k=0,\ldots,n}^{k\neq r} \Phi_k C_k h^{*-2k}(1+h^*)^{2k} + \Phi_r h^{*-(2r+1)}[C^* + |\,c_r\,| h^*(1+h^*)^{2r}]\,.$$

Therefore the set $\{u_\nu\}_{\nu=1,2,\ldots}$ is bounded in H_{A_0}, and the proof is completed as in the previous theorem.

Examples 3.12

1) $A_0 u = (x^4 u'')'' - (a_1(x)u')'$, $D(A_0) = C_0^\infty(a,\infty)$, $a > 0$,
$a_1(x) \geqslant -Cx^2$, $a \leqslant x < \infty$.

The spectrum of any self-adjoint extension of A_0 is discrete if and only if

$$\lim_{x\to\infty} x^{-3} \int_x^{x+hx} a_1(t)\,dt = \infty$$

for each fixed $h > 0$.

2) $A_0 u = (x^{2\alpha} u'')'' + a_0(x)u$, $\alpha \leqslant 2$, $D(A_0) = C_0^\infty(a,\infty)$, $a > 0$,
$a_0(x) \geqslant -C$, $a \leqslant x < \infty$.

We have the discreteness of the spectrum if and only if

$$\lim_{x\to\infty} x^{-(\alpha/2)} \int_x^{x+hx^{\alpha/2}} a_0(t)\,dt = \infty$$

for each fixed $h > 0$.

SECOND-ORDER DIFFERENTIAL OPERATORS

Next we consider necessary and sufficient conditions for the discreteness of the spectrum of second-order differential operators where the behaviour of the highest coefficient is not restricted by a power x^α.

Theorem 3.13
Let the coefficients of the Sturm-Liouville differential expression

$$a[\cdot] = -\frac{d}{dx}\,p(x)\frac{d}{dx} + q(x), \quad 0 \leqslant x < \infty,$$

satisfy the following conditions.

i) $p(x) \in W_2^1(0, X)$ and $q(x) \in L_2(0, X)$ for all $X > 0$.

ii) $p(x) > 0$, $0 \leqslant x < \infty$, and $\liminf\limits_{x\to\infty} q(x) > -\infty$.

iii) There are positive constants h_0, c_0 and C_0 such that

$$c_0\,p(x_1) \leqslant p(x_2) \leqslant C_0\,p(x_1) \tag{43}$$

for $0 \leqslant x_1 \leqslant x_2 \leqslant x_1 + h_0 p^{1/2}(x_1)$.

Then the spectrum of any self-adjoint extension of A_0,

$$A_0 u = a[u], \quad D(A_0) = C_0^\infty(0, \infty),$$

is discrete if and only if

$$\lim_{x\to\infty} \frac{1}{p^{1/2}(x)} \int_x^{x+hp^{1/2}(x)} q(t)\,dt = \infty \quad \text{for each fixed } h, 0 < h < 1. \tag{44}$$

Proof
Without loss of generality we assume that $q(x) \geqslant 0$ for $x \geqslant 1$. We choose any h, $0 < h < \min(h_0, 1)$ and, defining

$$x_{\nu+1} = x_\nu + hp^{1/2}(x), \quad \nu = 1, 2, \ldots, \quad x_1 = 0, \tag{45}$$

we obtain a sequence of intervals $\omega_\nu = [x_\nu, x_{\nu+1}]$ with $|\omega_\nu| = hp^{1/2}(x_\nu)$. The union of these intervals ω_ν, $\nu = 1, 2, \ldots$, is the half axis $[0, \infty)$. This is so because, in the opposite case, there is a finite ξ such that $x_\nu \to \xi$ as $\nu \to \infty$, and then we have the two contradictory situations that $|\omega_\nu| \to 0$ while, for all ν,

$$|\omega_\nu| \geqslant h \left(\min_{0 \leqslant x \leqslant \xi} p(x) \right)^{1/2}.$$

By Lemma 3.1 we have

$$\begin{aligned}(A_0 u, u) &= \sum_{\nu=1}^{\infty} \int_{\omega_\nu} (p(x)|u'|^2 + q(x)|u|^2)\,\mathrm{d}x \\ &\geqslant \sum_{\nu=1}^{\infty} \left(|\omega_\nu| \int_{\omega_\nu} \frac{\mathrm{d}x}{p(x)} + \frac{|\omega_\nu|}{\int_{\omega_\nu} q(x)\,\mathrm{d}x} \right)^{-1} \|u\|^2_{\omega_\nu},\end{aligned}$$

where $u \in C_0^\infty(1, \infty)$. Then, by (43) we obtain

$$\begin{aligned}(A_0 u, u) &\geqslant \sum_{\nu=1}^{\infty} \left\{ h^2 p(x_\nu) c_0^{-1} p^{-1}(x_\nu) + |\omega_\nu| \left(\left(\int_{\omega_\nu} q(x)\,\mathrm{d}x \right)^{-1} \right\}^{-1} \|u\|^2_{\omega_\nu} \right. \\ &= \sum_{\nu=1}^{\infty} \left\{ h^2 c_0^{-1} + hp^{1/2}(x_\nu) \left(\int_{x_\nu}^{x_\nu + hp^{1/2}(x_\nu)} q(x)\,\mathrm{d}x \right)^{-1} \right\}^{-1} \|u\|^2_{\omega_\nu}.\end{aligned} \tag{46}$$

If we now choose h so small that $h^2 c_0^{-1} \leqslant \epsilon$, where ϵ (>0) is arbitrary, then by (44) we have

$$h^2 c_0^{-1} + hp^{1/2}(x_\nu) \left(\int_{x_\nu}^{x_\nu + hp^{1/2}(x_\nu)} q(x)\,\mathrm{d}x \right)^{-1} \leqslant 2\epsilon$$

for $\nu \geqslant N(\epsilon)$. Hence from (46) we obtain

$$(A_0 u, u) \geqslant \sum_{\nu=1}^{\infty} (2\epsilon)^{-1} \|u\|^2_{\omega_\nu} = (2\epsilon)^{-1} \|u\|^2 \tag{47}$$

for functions $u(x) \in C_0^\infty(x_\epsilon, \infty)$, where $x_\epsilon = x_{N(\epsilon)}$. Consequently, if we decompose the operator A_0 at x_ϵ and define the operator $A_{(x_\epsilon, \infty),0}$, then we have

$$\sigma_e(\hat{A}_{(x_\epsilon, \infty)}) \cap (-\infty, (2\epsilon)^{-1}) = \emptyset,$$

where $\hat{A}_{(x_\epsilon, \infty)}$ denotes the Friedrichs extension of $A_{(x_\epsilon, \infty),0}$. Hence also

$$\sigma_e(\hat{A}) \cap (-\infty, (2\epsilon)^{-1}) = \emptyset$$

and, since ϵ is arbitrary, we obtain $\sigma_e(\hat{A}) = \emptyset$.

To prove the necessity of condition (44) for the discreteness of the spectrum we assume now that (44) does not hold. Then there are constants $h^* > 0$ and $C^* < \infty$ such that

$$\frac{1}{p^{1/2}(\xi_\nu)} \int_{\xi_\nu}^{\xi_\nu + h^* p^{1/2}(\xi_\nu)} q(t)\,\mathrm{d}t \leqslant C^*, \quad \nu = 1, 2, \ldots \tag{48}$$

for countable values ξ_ν, $\nu = 1, 2, \ldots$. Here we may assume that

$$\xi_\nu + h^* p^{1/2}(\xi_\nu) < \xi_{\nu+1}, \quad \nu = 1, 2, \ldots, \quad \xi_1 > 1. \tag{49}$$

Let $\varphi(x) \in C_0^\infty(-\infty, \infty)$ be a real-valued function with $\|\varphi\| = 1$, whose support is contained in the interval $(0, 1)$. Then the functions

$$u_\nu(x) = h^{*-(1/2)} p^{-(1/4)}(\xi_\nu) \varphi[h^{*-1} p^{-(1/2)}(\xi_\nu)(x - \xi_\nu)], \quad \nu = 1, 2, \ldots$$

have the properties

$$\|u_\nu\|^2 = h^{*-1} p^{-(1/2)}(\xi_\nu) \int_{\omega_\nu} \varphi^2 [h^{*-1} p^{-(1/2)}(\xi_\nu)(x - \xi_\nu)]\,\mathrm{d}x = 1$$

and

$$[u_\nu^{(k)}(x)]^2 \leqslant h^{*-(2k+1)} p^{-(k+1/2)}(\xi_\nu) \cdot \max_{0<t<1} [\varphi^{(k)}(t)]^2$$

$$\leqslant \frac{\Phi_k}{|\omega_\nu|^{2k+1}}, \quad k = 0, 1 \tag{50}$$

where

$$|\omega_\nu| = h^* p^{1/2}(\xi_\nu). \tag{51}$$

Now, A_0 is bounded from below and, to prove the boundedness of $\{u_\nu\}_{\nu=1,2,\ldots}$ in the energy space H_{A_0}, it is sufficient to estimate $(A_0 u_\nu, u_\nu)$. By (43), (48), (50), and (51) we obtain

$$(A_0 u_\nu, u_\nu) = \int_{\omega_\nu} (p(x)|u_\nu'|^2 + q(x)|u_\nu|^2)\,\mathrm{d}x$$

$$\leqslant \frac{\Phi_1}{|\omega_\nu|^3} \int_{\omega_\nu} p(x)\,\mathrm{d}x + \frac{\Phi_0}{|\omega_\nu|} \int_{\omega_\nu} q(x)\,\mathrm{d}x$$

$$\leqslant \frac{\Phi_1}{|\omega_\nu|^2} C_0 p(\xi_\nu) + \Phi_0 h^{*-1} C^*$$

$$= \Phi_1 h^{*-2} C_0 + \Phi_0 h^{*-1} C^*.$$

Therefore by Theorem 1.2 the essential spectrum of A is not empty, and Theorem 3.13 is proved.

Note 3.14

From the first part of the proof of Theorem 3.13 we see that the spectrum is discrete if (44) is satisfied and $p(x)$ satisfies the condition

$$c_0 p(x_1) \leqslant p(x_2), \quad x_1 \leqslant x_2 \leqslant x_1 + h_0 p^{1/2}(x_1),$$
$$x_1, x_2 \in (0, \infty), \quad c_0 > 0 \,. \tag{52}$$

Every increasing function $p(x)$ satisfies (52) for instance.

One can easily prove that the power function x^{α}, $\alpha \leqslant 2$, satisfies (43) as also does

$$p(x) = Ce^{-x^{\beta}}, \quad 0 \leqslant \beta < \infty \,.$$

Note 3.15

Without giving the proof we mention the following result for the special operator

$$A_0 u = (-1)^n (a_n(x) u^{(n)})^{(n)}, \quad D(A_0) = C_0^{\infty}(0, \infty), \quad a_n(x) > 0, \quad x \geqslant 0.$$

The spectrum of any self-adjoint extension of A_0 is discrete if and only if

$$\lim_{x \to \infty} x^{2n-1} \int_x^{\infty} \frac{\mathrm{d}t}{a_n(t)} = 0 \,.$$

The sufficiency of this condition was proved by Tkachenko (see [15] §34) and the necessity by Lewis [32].

In the next theorem a necessary and sufficient condition for the discreteness of the spectrum is given in the case that the basic interval is finite.

Theorem 3.16

Let the coefficients of the Sturm-Liouville differential expression

$$a[\cdot] = -\frac{d}{\mathrm{d}x} p(x) \frac{d}{\mathrm{d}x} + q(x), \quad 0 < x \leqslant 1 \,.$$

satisfy the following conditions.

i) $p(x) \in W_2^1(X, 1)$ and $q(x) \in L_2(X, 1)$ for all $X, 0 < X < 1$.

ii) There is a constant $C_p > 0$ such that

$$0 < p(x) \leqslant C_p x^2, \quad q(x) \geqslant 0, \quad 0 < x \leqslant 1 \,.$$

iii) There exist positive constants h_0, c_0, C_0 such that

$$c_0 p(x_1) \leqslant p(x_2) \leqslant C_0 p(x_1), \quad x_1 \leqslant x_2 \leqslant x_1 + h_0 p^{1/2}(x_1),$$
$$0 < x_i \leqslant 1, i = 1, 2 .$$

Then the spectrum of any self-adjoint extension of A_0, $A_0 u = a[u]$, $D(A_0) = C_0^\infty(0, 1)$, is discrete if and only if

$$\lim_{x \downarrow 0} \frac{1}{p^{1/2}(x)} \int_x^{x+hp^{1/2}(x)} q(t)\, dt = \infty \tag{53}$$

for each $h, 0 < h < \min(h_0, C_p^{-(1/2)})$.

Proof
We define a sequence x_ν by

$$x_{\nu+1} = x_\nu - hp^{1/2}(x_\nu), \quad 0 < h < \min(h_0, C_p^{-(1/2)}),$$
$$\nu = 1, 2, \ldots, \quad x_1 = 1,$$

and we have $x_\nu \to 0$ as $\nu \to \infty$. We still have (46) for $u \in C_0^\infty(0, 1)$ where now $\omega_\nu = [x_{\nu+1}, x_\nu]$. As in the proof of Theorem 3.13 for a given $\epsilon > 0$ we find a value x_ϵ, $0 < x_\epsilon < 1$, such that

$$(A_0 u, u)_{(0,1)} \geqslant (2\epsilon)^{-1} \| u \|^2_{(0,1)}, \quad u \in C_0^\infty(0, x_\epsilon) .$$

Then as before this gives $\sigma_e(\hat{A}) = \emptyset$. The necessity of (53) for discreteness of the spectrum also follows as in the proof of Theorem 3.13, and Theorem 3.16 is proved.

We also indicate that in the case

$$p(x) \geqslant cx^\alpha, \quad 0 < x \leqslant 1, \quad \alpha < 2, \quad c > 0 ,$$

(53) is no longer necessary for the discreteness of the spectrum. Indeed, in the case $q(x) \equiv 0$, the spectrum is entirely discrete as can be seen by using Courant's variational principle and comparing with the example

$$B_0 u = -K(x^2 u')', \quad K > 0, \quad D(B_0) = C_0^\infty(0, 1) .$$

Finally in this section we mention a sufficient condition for the discreteness of the spectrum by Friedrichs [14]. The spectrum of any self-adjoint extension of A_0,

$$A_0 u = -(p(x)u')' + q(x)u, \quad p(x) > 0, \quad 0 \leqslant x < \infty,$$
$$D(A_0) = C_0^\infty(0, \infty),$$

is discrete when

$$\lim_{x \to \infty} \left[q(x) + \left\{ 4p(x) \left(\int_0^x p^{-1}(t)\,\mathrm{d}t \right)^2 \right\}^{-1} \right] = \infty .$$

There are series of other sufficient criteria for the discreteness of the spectrum which are not dealt with here. A summary of such criteria can be found in [8].

THE INFIMUM OF THE SPECTRUM OF PERIODIC DIFFERENTIAL OPERATORS

We close this chapter with some remarks on the spectrum of differential operators with periodic coefficients. Here the basic interval is $(-\infty, \infty)$ and all coefficients $a_k(x)$, $0 \leqslant k \leqslant n$, have the same period. It is well known ([8], [10]) that such differential operators possess no eigenvalues. In the next theorem the position of the lowest point of the spectrum is estimated by means of Lemma 3.1. We denote the interval of periodicity by ω and its length by $|\omega|$, thus $\omega = (0, |\omega|)$.

Theorem 3.17
Let the coefficients $a_k(x)$ satisfy the following conditions.

i) $a_k(x + |\omega|) = a_k(x)$, $|\omega| > 0$, $-\infty < x < \infty$, $0 \leqslant k \leqslant n$,

ii) $a_k(x) \in W_{2,\mathrm{loc}}^k(-\infty, \infty)$, $0 \leqslant k \leqslant n$,

$\inf_{x \in \omega} a_n(x) > 0$, $a_k(x) \geqslant 0$, $0 \leqslant k \leqslant n-1$.

Then the operator A_0,

$$A_0 u = a[u], \quad D(A_0) = C_0^\infty(-\infty, \infty),$$

is essentially self-adjoint. If Λ denotes the infimum of the spectrum of the closure A of A_0 and we define

$$\mu_k = \frac{1}{|\omega|} \int_\omega a_k(x)\,\mathrm{d}x, \; 0 \leqslant k \leqslant n-1, \quad \rho_n = \left(\frac{1}{|\omega|} \int_\omega \frac{\mathrm{d}x}{a_n(x)} \right)^{-1},$$

then

$$\Lambda \geqslant \left(|\omega|^{2n} \rho_n^{-1} + \sum_{k=0}^{n-1} |\omega|^{2k} \mu_k^{-1} \right)^{-1} .$$

Proof
The essential self-adjointness of A_0 follows immediately from Theorem 2.4. If the x-axis is decomposed in countable intervals $\omega_\nu = (\nu|\omega|, (\nu+1)|\omega|)$, $\nu = 0, \pm 1, \pm 2, \ldots$, then by Lemma 3.1 we have

$$(A_0 u, u) \geqslant \sum_{\nu=-\infty}^{\infty} \left(|\omega|^{2n} \rho_n^{-1} + \sum_{k=0}^{n-1} |\omega|^{2k} \mu_k^{-1} \right)^{-1} \| u \|^2_{\omega_\nu}$$

$$\geqslant \left(|\omega|^{2n} \rho_n^{-1} + \sum_{k=0}^{n-1} |\omega|^{2k} \mu_k^{-1} \right)^{-1} \| u \|^2, \quad u \in C_0^\infty(-\infty, \infty) .$$

This is also true for the closure A, and with this Theorem 3.17 is proved.
In the case $n = 1$ the following theorem can be proved.

Theorem 3.18 [43]
Let

i) $a_k(x + |\omega|) = a_k(x)$, $|\omega| > 0$, $-\infty < x < \infty$, $k = 0, 1$,
ii) $a_k(x) \in W^k_{2,\mathrm{loc}}(-\infty, \infty)$, $k = 0, 1$,
$\inf_{x \in \omega} a_1(x) > 0$, $a_0(x) \geqslant 0$, $0 \leqslant x < |\omega|$.

Then for $\Lambda = \inf \sigma(A)$ the estimate

$$\frac{\mu}{1 + \eta\mu} \leqslant \Lambda \leqslant \mu \tag{54}$$

holds where

$$\mu = \frac{1}{|\omega|} \int_\omega a_0(x)\,\mathrm{d}x, \quad \eta = \inf_{\xi \in \omega} \max \left(\int_\xi^{\xi+|\omega|} \frac{x - \xi}{a_1(x)}\,\mathrm{d}x, \int_\xi^{\xi+|\omega|} \frac{\xi + |\omega| - x}{a_1(x)}\,\mathrm{d}x \right) .$$

Proof
If the real-valued function $\varphi(x) \in C_0^\infty(-\infty, \infty)$ possesses a zero at $x_0 \in [\xi, \xi + |\omega|]$, then by the Schwarz inequality we have

$$\varphi^2(x) = \left(\int_{x_0}^x \varphi'(t)\,\mathrm{d}t \right)^2 \leqslant \left| \int_{x_0}^x \frac{\mathrm{d}t}{a_1(t)} \right| \cdot \int_\xi^{\xi+|\omega|} a_1(t)[\varphi'(t)]^2\,\mathrm{d}t$$

$$x \in [\xi, \xi + |\omega|] ,$$

and, using

$$\int_{\xi}^{\xi+|\omega|}\int_{x_0}^{x}\frac{dt}{a_1(t)}\,dx=\int_{\xi}^{x_0}\frac{x-\xi}{a_1(x)}\,dx+\int_{x_0}^{\xi+|\omega|}\frac{\xi+|\omega|-x}{a_1(x)}\,dx$$

$$\leqslant \max\left(\int_{\xi}^{\xi+|\omega|}\frac{x-\xi}{a_1(x)}\,dx,\quad \int_{\xi}^{\xi+|\omega|}\frac{\xi+|\omega|-x}{a_1(x)}\,dx\right)=\eta_\xi\,,$$

say, we obtain

$$\int_{\xi}^{\xi+|\omega|}\varphi^2(x)\,dx\leqslant\eta_\xi\int_{\xi}^{\xi+|\omega|}a_1(x)[\varphi'(x)]^2\,dx\,.$$

Thus, if $\xi\in[0,|\omega|]$ is the point where η_ξ is minimal we have

$$\int_{\xi}^{\xi+|\omega|}\varphi^2(x)\,dx\leqslant\eta\int_{\xi}^{\xi+|\omega|}a_1(x)[\varphi'(x)]^2\,dx \tag{55}$$

and then with a similar argument to that used in the proof of Lemma 2.15 we obtain

$$\frac{\mu}{1+\eta\mu}\,\|u\|^2_{(\xi,\xi+|\omega|)}\leqslant\int_{\xi}^{\xi+|\omega|}(a_1(x)|u'|^2+a_0(x)|u|^2)\,dx,$$

$$u(x)\in C_0^\infty[\xi,\xi+|\omega|]. \tag{56}$$

If $(-\infty,\infty)$ is now expressed as the union of disjoint intervals which arise from the interval $[\xi,\xi+|\omega|)$ by translations, then by addition from (56) we obtain

$$\frac{\mu}{1+\eta\mu}\,\|u\|^2_{(-\infty,\infty)}\leqslant\int_{-\infty}^{\infty}(a_1(x)|u'|^2+a_0(x)|u|^2)\,dx,$$

$$u\in C_0^\infty(-\infty,\infty)\,.$$

The left-hand inequality in (54) now follows.

The other part $\Lambda\leqslant\mu$ follows from

$$\Lambda=\inf_{\|u\|>0}\left(\|u\|^{-2}\int_{-\infty}^{\infty}(a_1(x)|u'|^2+a_0(x)|u|^2)\,dx\right),$$

$$u\in C_0^\infty(-\infty,\infty),$$

if we use functions $u_N(x)$ which are generated from the functions

$$U_N(x) = \begin{cases} 1, |x| \leqslant N, \\ 0, |x| > N, \end{cases}$$

by averaging with a fixed radius and then let N tend to infinity. Theorem 3.18 is now proved.

By decomposition of the interval of periodicity we can improve the lower bound for Λ in (54). Defining

$$\eta_\nu = \max\left(\int_{\xi_\nu}^{\xi_{\nu+1}} \frac{x-\xi_\nu}{a_1(x)}\,\mathrm{d}x,\ \int_{\xi_\nu}^{\xi_{\nu+1}} \frac{\xi_{\nu+1}-x}{a_1(x)}\,\mathrm{d}x\right),$$

$$\xi = \xi_0 < \xi_1 < \cdots < \xi_n = \xi + |\omega|,$$

$$\mu_\nu = \frac{1}{\xi_{\nu+1}-\xi_\nu}\int_{\xi_\nu}^{\xi_{\nu+1}} a_0(x)\,\mathrm{d}x,\quad \sigma_\nu = \frac{\mu_\nu}{1+\eta_\nu\mu_\nu},$$

$$\nu = 0, 1, \ldots, n-1,$$

$$s(\xi, z_n) = \min_{0\leqslant\nu\leqslant n-1} \sigma_\nu,\quad S = \sup_{(\xi, z_n)} s(\xi, z_n),$$

where z_n denotes the decomposition $z_n = \{\xi = \xi_0, \xi_1, \ldots, \xi_n\}$, we obviously have $S \leqslant \Lambda$ again. Since, for instance, $a_0(x)$ has the period $|\omega|$ there exist points $\xi = \xi_0$, $\xi_1 = \xi + (|\omega|/2)$, $\xi_2 = \xi + |\omega|$ such that $\mu_0 = \mu_1 = \mu$, and then we obtain

$$\frac{\mu}{1+\tilde{\eta}\mu} \leqslant \Lambda \tag{57}$$

where

$$\tilde{\eta} = \max\left(\int_{\xi}^{\xi+(|\omega|/2)} \frac{x-\xi}{a_1(x)}\,\mathrm{d}x,\ \int_{\xi}^{\xi+(|\omega|/2)} \frac{\xi+(|\omega|/2)-x}{a_1(x)}\,\mathrm{d}x,\right.$$
$$\left.\int_{\xi+(|\omega|/2)}^{\xi+|\omega|} \frac{x-\xi-(|\omega|/2)}{a_1(x)}\,\mathrm{d}x,\ \int_{\xi+(|\omega|/2)}^{\xi+|\omega|} \frac{\xi+|\omega|-x}{a_1(x)}\,\mathrm{d}x\right).$$

In the special case $a_1(x) \equiv 1$ this gives $\tilde{\eta} = |\omega|^2/8$. To obtain a smaller η $(= |\omega|^2/\pi^2)$, we prove

Lemma 3.19
If the function $\varphi(x) \in C^1[0, a]$ has a zero on $[0, a]$, then

$$\int_0^a |\varphi'|^2\,\mathrm{d}x \geqslant \frac{\pi^2}{4a^2}\int_0^a |\varphi|^2\,\mathrm{d}x. \tag{58}$$

Proof

We first consider functions which have a zero at $x = 0$. The minimum of the quadratic form

$$\int_0^a |\varphi'|^2 \mathrm{d}x, \quad \varphi(x) \in W_2^1(0, a), \quad \varphi(0) = 0, \quad \|\varphi\|_{(0,a)} = 1,$$

is realized by the eigenfunction

$$\varphi(x) = \sqrt{\frac{2}{a}} \sin\left(\frac{\pi}{2a}x\right), \quad 0 \leqslant x \leqslant a,$$

of the self-adjoint operator which is defined by the form

$$\int_0^a \varphi' \overline{\psi'} \mathrm{d}x, \quad \varphi, \psi \in W_2^1(0, a), \quad \varphi(0) = \psi(0) = 0,$$

(see Chapter 4, section 1). Hence we have (58) when $\varphi(0) = 0$ (or $\varphi(a) = 0$). Now it is easily seen that (58) is true when the zero lies in $(0, a)$, if we divide $(0, a)$ into two parts by the zero.

As with (55) and (56), from (58) we obtain

$$\frac{\pi^2 \mu}{\pi^2 + |\omega|^2 \mu} \|u\|^2_{(\xi_\nu, \xi_\nu + (|\omega|/2))} \leqslant \int_{\xi_\nu}^{\xi_\nu + (|\omega|/2)} (|u'|^2 + a_0(x)|u|^2) \mathrm{d}x,$$

$$u \in C^\infty[\xi_0, \xi_2], \nu = 0, 1,$$

where ξ_0, ξ_1, ξ_2 are the points already defined in obtaining (57). Thus we have

Theorem 3.20

Let $a_1(x) \equiv 1$ and $a_0(x) \geqslant 0, 0 \leqslant x < |\omega|$. Then

$$\frac{\pi^2 \mu}{\pi^2 + |\omega|^2 \mu} \leqslant \Lambda \leqslant \mu,$$

Note 3.21

Another possible improvement of the lower bound for Λ is given by writing

$$a_0(x) = a_0^+(x) + a_0^-(x), \quad a_0^+(x) = \max(a_0(x), 0),$$

$$a_0^-(x) = \min(a_0(x), 0),$$

and defining

$$\mu^+ = \frac{1}{|\omega|}\int_0^\omega a_0^+(x)\,dx, \quad \mu^- = \frac{1}{|\omega|}\int_0^\omega a_0^-(x)\,dx\,.$$

Then the decomposition

$$\int_{-\infty}^{\infty}\left(\frac{a_1(x)}{\alpha}|u'|^2 + a_0^+(x)|u|^2\right)dx + \int_{-\infty}^{\infty}\left(\frac{a_1(x)}{\beta}|u'|^2 + a_0^-(x)|u|^2\right)dx,$$

$$u \in C_0^\infty(-\infty, \infty), \quad \frac{1}{\alpha}+\frac{1}{\beta} = 1, \quad \alpha, \beta > 1\,,$$

of the quadratic form of A_0 leads to

$$\frac{\mu^+}{1+\alpha\eta\mu^+} + \frac{\mu^-}{1+\beta\eta\mu^-} \leqslant \Lambda \leqslant \mu, \quad \beta < |\mu^-|^{-1}\eta^{-1}\,.^\dagger$$

If we choose the values

$$\alpha = \frac{\mu^+ + |\mu^-|}{\mu^+ + 2\eta\mu^+\mu^-}, \quad \beta = \frac{\mu^+ + |\mu^-|}{|\mu^-| + 2\eta\mu^+|\mu^-|},$$

granted that

$$\mu^+ > 0, \quad |\mu^-| > 0, \quad |\mu^-| < \frac{1}{2\eta}, \quad \mu^+ + \mu^- = \mu > -\eta^{-1}\,,$$

then we obtain

$$\frac{\mu - 4\eta\mu^+|\mu^-|}{1+\eta\mu} \leqslant \Lambda \leqslant \mu\,.$$

In the case of $a_1(x) \equiv 1$ we again take $\eta = \omega^2\pi^{-2}$ and obtain

$$\frac{\mu\pi^2 - 4\omega^2\mu^+|\mu^-|}{\pi^2 + \omega^2\mu} \leqslant \Lambda \leqslant \mu\,, \quad |\mu^-| < \frac{\pi^2}{2\omega^2}, \quad \mu > -\frac{\pi^2}{\omega^2}\,.$$

† Use the corresponding generalization of the estimate (73) (Chapter 2); there we have $\alpha = \beta = 2$.

Other estimates for Λ have been given by Kato [25] and Eastham [10]; they are

$$\mu - \frac{1}{16}\left(\int_0^{|\omega|} |a_0(x) - \mu| \, dx\right)^2 \leqslant \Lambda \leqslant \mu, \quad a_1(x) \equiv 1,$$

and

$$\mu - \frac{1}{16M}\left(\int_0^{|\omega|} |a_0(x) - \mu| \, dx\right)^2 \leqslant \Lambda \leqslant \mu, \quad M = \inf a_1(x),$$

respectively. Here we refer to [10] for a general account of the spectrum of periodic differential operators of the second order.

Chapter 4

Continuous Spectra

In the previous two chapters the essential spectrum was examined. The second chapter dealt with the stability of the essential spectrum under perturbations and the last chapter dealt with the special case $\sigma_e(A) = \emptyset$. There, the boundary conditions which determine the self-adjoint extensions of $A_0, D(A_0) = C_0^\infty(0, \infty)$, were not involved because the essential spectrum is the same for all such extensions. In the present chapter, on the other hand, boundary conditions are significant because the existence and position of the eigenvalues depend not only on the coefficients but also on the boundary conditions. Accordingly the continuity of the spectrum on intervals is influenced by the boundary conditions. In what follows we give conditions on the coefficients $a_k(x)$ and on the boundary behaviour of functions belonging to the domains of the operators so that the spectrum of the operator in question is continuous or possesses no eigenvalues on certain intervals of the λ-axis.

SELF-ADJOINT OPERATORS AND BOUNDARY CONDITIONS

We now consider self-adjoint differential operators of order $2n$ which are generated by sesquilinear forms associated with the differential expression

$$a[\cdot] = \sum_{k=0}^{n} (-1)^k \frac{d^k}{dx} a_k(x) \frac{d^k}{dx^k}, \quad 0 < x < \infty.$$

The following conditions on the coefficients are required.

i) $a_k(x) \in W_{2,\mathrm{loc}}^k(0, \infty)$ real-valued, $0 \leqslant k \leqslant n, a_n(x) > 0, 0 < x < \infty$. (1)

ii) There exist $\delta > 0$ and l, $0 \leqslant l < n$, such that

$$a_k(x) \geqslant 0, \quad 0 < x < \infty, \quad l + 1 \leqslant k \leqslant n,$$

$$\int_0^\delta a_{l+1}^{-1}(x)\, dx < \infty, \quad \int_0^1 |a_j(x)|\, dx < \infty, \quad 0 \leqslant j \leqslant l.^\dagger \tag{2}$$

†If (2) is satisfied in the case $l + 1 = n$, then according to [46] the boundary point $x = 0$ is said to be regular with respect to $a[\cdot]$. Otherwise the boundary point is said to be singular.

iii) There are α, $-\infty < \alpha \leqslant 2$, and k^*, $l+1 \leqslant k^* \leqslant n$, such that

$$\liminf_{x\to\infty} a_{k^*}(x)x^{-\alpha k^*} > 0, \quad \liminf_{x\to\infty} a_j(x)x^{-\alpha j} > -\infty, \quad 0 \leqslant j \leqslant l\,. \tag{3}$$

The condition (iii) is absent if no $a_j(x)$ is negative for large x.

Let

$$(\alpha) = \begin{pmatrix} \alpha_{00} & \dots & \alpha_{0l} & \alpha_{0,l+1} & \dots & \alpha_{0,2l+1} \\ \vdots & & \vdots & \vdots & & \vdots \\ \alpha_{l0} & \dots & \alpha_{ll} & \alpha_{l,l+1} & \dots & \alpha_{l,2l+1} \end{pmatrix},$$

$$\alpha_{ij} \text{ complex}, \; 0 \leqslant i \leqslant l, \quad 0 \leqslant j \leqslant 2l+1\,,$$

be a matrix with $\alpha_{ij} = \overline{\alpha_{ji}}$, $0 \leqslant i, j \leqslant l$. Then the sesquilinear form

$$a_{(\alpha)}[u, v] = \sum_{k=0}^{n} \int_0^\infty a_k(x)u^{(k)}(x)\bar{v}^{(k)}(x)\,\mathrm{d}x + \sum_{i,j=0}^{l} \alpha_{ij}u^{(i)}(0)\bar{v}^{(j)}(0)$$

with the domain

$$D(a_{(\alpha)}) = \{u \mid u \in C^n[0, \infty), u(x) = 0, x > X_u\;;\; u^{(l+1)}(x) = 0,$$
$$0 \leqslant x < x_u;\; \sum_{j=0}^{l} \alpha_{i,l+1+j}u^{(j)}(0) = 0, i = 0, \dots, l\}$$

is symmetric.

Lemma 4.1

Let (1), (2) and (3) hold. Then the form $a_{(\alpha)}[u, v]$, $D(a_{(\alpha)})$, is semibounded from below and closeable.

Proof

First we show that the form

$$a^+[u, v] = \sum_{k=0}^{n} \int_0^\infty a_k^+(x)u^{(k)}(x)\bar{v}^{(k)}(x)\,\mathrm{d}x\,,$$

$$a_k^+(x) = \max(a_k(x), 0), \quad D(a^+) = D(a_{(\alpha)})\,,$$

which is semibounded from below, is closeable. By means of a^+ the norm

$$\| u \|_{a^+} = \left(\sum_{k=0}^{n} \int_0^\infty a_k^+(x)|\, u^{(k)}\,|^2\,\mathrm{d}x + \| u \|^2 \right)^{1/2}$$

is defined and, according to Theorem 1.3, we have to show that $\| u_\nu \| \to 0$, $\nu \to \infty$, implies that $\| u_\nu \|_{a^+} \to 0$, $\nu \to \infty$, where $\{u_\nu\}_{\nu=1,2,\ldots}$, $u_\nu \in D(a^+)$, is a $\| \cdot \|_{a^+}$-Cauchy sequence. So we assume that

$$\| u_\nu - u_{\nu'} \|_{a^+} \to 0, \quad \nu, \nu' \to \infty, \quad \text{and} \quad \| u_\nu \| \to 0, \quad \nu \to \infty .$$

The set

$$\Xi_\epsilon = \{x \mid \epsilon < a_n(x) \text{ and } a_k^+(x) < \epsilon^{-1}, 1 \leqslant k \leqslant n\}, \quad 0 < \epsilon < 1 ,$$

is open because of the continuity of $a_k^+(x)$, $1 \leqslant k \leqslant n$, which follows from (1). Let ϵ_1 be chosen so small that $x = 1$ belongs to Ξ_ϵ for all ϵ with $0 < \epsilon \leqslant \epsilon_1$, and let Ω_ϵ be the component of Ξ_ϵ that contains the point $x = 1$. From the estimate

$$\begin{aligned} &\epsilon \| u^{(n)} \|^2_{\Omega_\epsilon} + \| u \|^2_{\Omega_\epsilon} \leqslant \| u \|^2_{a^+, \Omega_\epsilon} \\ &\leqslant \epsilon^{-1} \| u \|^2_{W_2^n(\Omega_\epsilon)} + \| u \|^2_{\Omega_\epsilon} \leqslant 2\epsilon^{-1} \| u^{(n)} \|^2_{\Omega_\epsilon} + C_\epsilon \| u \|^2_{\Omega_\epsilon} \end{aligned} \tag{4}$$

it follows that $\{u_\nu\}_{\nu=1,2,\ldots}$ is a Cauchy sequence in the Sobolev space $W_2^n(\Omega_\epsilon)$. Since $\{u_\nu\}_{\nu=1,2,\ldots}$ tends to the zero-element in $L_2(\Omega_\epsilon)$ it follows from

$$\int_{\Omega_\epsilon} \varphi u_\nu^{(n)} \, dx = (-1)^n \int_{\Omega_\epsilon} \varphi^{(n)} u_\nu \, dx, \quad \nu = 1, 2, \ldots, \quad \varphi \in C_0^\infty(\Omega_\epsilon),$$

that $\{u_\nu^{(n)}\}_{\nu=1,2,\ldots}$ also tends to the zero-element in $L_2(\Omega_\epsilon)$. Therefore by (4) we have

$$\| u_\nu \|_{a^+, \Omega_\epsilon} \to 0, \quad \nu \to \infty . \tag{5}$$

If we now assume, as a proof by contradiction, that there exists a subsequence $\{u_{\nu_j}\}_{j=1,2,\ldots}$ of $\{u_\nu\}_{\nu=1,2\ldots}$ such that

$$\| u_{\nu_j} \|_{a^+} \geqslant \eta > 0, \quad j = 1, 2, \ldots,$$

then by (5) we have

$$\| u_{\nu_j} \|_{a^+, \omega_\epsilon} \geqslant \tfrac{1}{2}\eta > 0, \quad j \geqslant j_1(\epsilon), \tag{6}$$

where $\omega_\epsilon = (0, \infty) \setminus \Omega_\epsilon$ and $j_1(\epsilon)$ is chosen sufficiently large. Now

$$\| u_{\nu_j} \|_{a^+, \omega_\epsilon} \geqslant \| u_{\nu_i} \|_{a^+, \omega_\epsilon} - \| u_{\nu_i} - u_{\nu_j} \|_{a^+, \omega_\epsilon} \tag{7}$$

and the second term on the right is not larger than $\frac{1}{4}\eta$ when $i, j \geqslant j_2$. Then, if for any ϵ, $0 < \epsilon \leqslant \epsilon_1$, we choose $i \geqslant \max(j_1(\epsilon), j_2)$, it follows from (6) and (7) that

$$\| u_{\nu_j} \|_{a^+,\omega_\epsilon} \geqslant \tfrac{1}{4}\eta, \quad j \geqslant j_2,$$

with j_2 independent of ϵ. This leads to a contradiction, if a function $u_{\nu_j}(x)$, $j \geqslant j_2$, is fixed and ϵ tends to zero.

We now show that the form

$$a_{(\alpha)}[u, v] = a^+[u, v] + a^-[u, v] + \sum_{i,j=0}^{l} \alpha_{ij} u^{(i)}(0)\, \bar{v}^{(j)}(0),$$
$$u, v \in D(a_{(\alpha)}),$$

is semibounded from below and closable. To do this, we prove that

$$\left| a^-[u, u] + \sum_{i,j=0}^{l} \alpha_{ij} u^{(i)}(0) \bar{u}^{(j)}(0) \right| \leqslant c_1 a^+[u, u] + c_2 \| u \|^2, \tag{8}$$
$$0 < c_1 < 1, \quad 0 < c_2 < \infty, \quad u \in D(a^+),$$

and then we can quote Theorem 1.4. To establish (8), we prove that for any ϵ there are constants C_ϵ such that

$$\left| \sum_{i,j=0}^{l} \alpha_{ij} u^{(i)}(0) \bar{u}^{(j)}(0) \right| \leqslant \epsilon a^+[u, u] + C_\epsilon \| u \|^2, \quad u \in D(a^+), \tag{9}$$

and

$$| a^-[u, u] | \leqslant \epsilon a^+[u, u] + C_\epsilon \| u \|^2, \quad u \in D(a^+). \tag{10}$$

We begin by proving that

$$\max_{0 \leqslant x \leqslant x_0} | u^{(i)}(x) |^2 \leqslant \epsilon a^+[u, u] + C_{\epsilon, x_0} \| u \|^2, \quad 0 \leqslant i \leqslant l, u \in D(a^+), \tag{11}$$

where x_0, $0 < x_0 < \infty$, and $\epsilon > 0$ can be chosen arbitrarily. Let the maximum of $| u^{(i)}(x) |$ on $[0, x_0]$ occur at ξ_i. We first assume that $0 \leqslant \xi_1 \leqslant \frac{1}{2}\delta$. If $\xi_i < \xi \leqslant \delta$ and the estimate (A.2) in Lemma A.1 is used, then

$$| u^{(i)}(\xi_i) - u^{(i)}(\xi) | \leqslant \int_{\xi_i}^{\xi} | u^{(i+1)} |\, dx \leqslant \int_{\xi_i}^{\xi} | u^{(l+1)} |\, dx + C_\xi \int_{\xi_i}^{\xi} | u |\, dx$$

$$\leqslant \left(\int_{\xi_i}^{\xi} a_{l+1}^{-1}(x)\, dx \right)^{1/2} \left(\int_{\xi_i}^{\xi} a_{l+1}(x) | u^{(l+1)} |^2\, dx \right)^{1/2} + C_\xi \| u \|_{(0,\delta)} .$$

We fix any ϵ_1 where

$$0 < \epsilon_1 \leqslant \int_{\delta/2}^{\delta} a_{l+1}^{-1}(x)\,\mathrm{d}x \,.$$

Then for ξ determined by

$$\int_{\xi_i}^{\xi} a_{l+1}^{-1}(x)\,\mathrm{d}x = \epsilon_1$$

we have $\xi_{\epsilon_1} \leqslant \xi \leqslant \delta$ where ξ_{ϵ_1} is defined by

$$\int_{0}^{\xi_{\epsilon_1}} a_{l+1}^{-1}(x)\,\mathrm{d}x = \epsilon_1 \,.$$

Here, ξ_{ϵ_1} does not depend on u. Hence we obtain

$$|u^{(i)}(\xi_i)|^2 \leqslant 3\epsilon_1 a^+[u,u] + C_{\epsilon_1}\|u\|^2 + 3|u^{(i)}(\xi)|^2 \,,$$

and, by (A.9), it follows that

$$\begin{aligned}
|u^{(i)}(\xi_i)|^2 &\leqslant 3\epsilon_1 a^+[u,u] + C_{\epsilon_1}\|u\|^2 + \epsilon_2\|u^{(n)}\|^2_{(\xi,\xi+1)} + C_{\epsilon_2}\|u\|^2_{(\xi,\xi+1)} \\
&\leqslant 3\epsilon_1 a^+[u,u] + \epsilon_2\left(\min_{\xi\leqslant x\leqslant \xi+1} a_n(x)\right)^{-1}\int_{\xi}^{\xi+1} a_n(x)|u^{(n)}|^2\,\mathrm{d}x + C_{\epsilon_1,\epsilon_2}\|u\|^2 \\
&\leqslant \left\{3\epsilon_1 + \epsilon_2\left(\min_{\xi_{\epsilon_1}\leqslant x\leqslant \delta+1} a_n(x)\right)^{-1}\right\} a^+[u,u] + C_{\epsilon_1,\epsilon_2}\|u\|^2.
\end{aligned}$$

Now (11) follows if first ϵ_1 and then ϵ_2 are chosen sufficiently small.

In the case $\frac{1}{2}\delta \leqslant \xi_i \leqslant x_0$ we use the inequality

$$\begin{aligned}
|u^{(i)}(\xi_i)|^2 &\leqslant \epsilon_3\|u^{(n)}\|^2_{(\delta/2,x_0)} + C_{\epsilon_3}\|u\|^2_{(\delta/2,x_0)} \\
&\leqslant \epsilon_3\left(\max_{\delta/2\leqslant x\leqslant x_0} a_n^{-1}(x)\right)\int_{\delta/2}^{x_0} a_n(x)\,|u^{(n)}|^2\,\mathrm{d}x + C_{\epsilon_3}\|u\|^2 \\
&\leqslant \epsilon_3\left(\max_{\delta/2\leqslant x\leqslant x_0} a_n^{-1}(x)\right) a^+[u,u] + C_{\epsilon_3}\|u\|^2 \,.
\end{aligned}$$

Now (11) follows again, with

$$\epsilon = \epsilon_3\left(\max_{\delta/2\leqslant x\leqslant x_0} a_n^{-1}(x)\right) \,.$$

From (11) we obtain (9) immediately.

To prove (10) we show that for any $\epsilon > 0$ there exists a constant C_ϵ such that

$$\int_0^\infty |a_j^-(x)||u^{(j)}|^2\,dx \leqslant \epsilon a^+[u,u] + C_\epsilon \|u\|^2, \quad u \in D(a^+), \quad 0 \leqslant j \leqslant l. \tag{12}$$

From (11) and (2) we obtain

$$\int_0^{x_0} |a_j^-(x)||u^{(j)}|^2\,dx \leqslant \left(\max_{0\leqslant x\leqslant x_0} |u^{(j)}(x)|^2\right)\int_0^{x_0} |a_j^-(x)|\,dx$$
$$\leqslant \epsilon a^+[u,u] + C_\epsilon \|u\|^2, \quad 0 \leqslant j \leqslant l. \tag{13}$$

To estimate

$$\int_{x_0}^\infty |a_j^-(x)||u^{(j)}|^2\,dx, \quad 0 \leqslant j \leqslant l,$$

we use (3) and (A.19). If x_0 is chosen so large that

$$a_j^-(x) \leqslant -Cx^{\alpha j}, \quad C>0, \quad 0\leqslant j\leqslant l, \quad a_{k^*}(x) \geqslant Cx^{\alpha k^*}, \quad C>0, \quad x \geqslant x_0$$

then we obtain

$$\int_{x_0}^\infty |a_j^-(x)||u^{(j)}|^2\,dx \leqslant C\int_{x_0}^\infty x^{\alpha j}|u^{(j)}|^2\,dx$$
$$\leqslant C\epsilon\int_{x_0}^\infty x^{\alpha k^*}|u^{(k^*)}|^2\,dx + CC_{x_0,\epsilon}\int_{x_0}^\infty |u|^2\,dx \tag{14}$$
$$\leqslant \epsilon'\int_{x_0}^\infty a_{k^*}(x)|u^{(k^*)}|^2\,dx + C_{x_0,\epsilon'}\int_{x_0}^\infty |u|^2\,dx \leqslant \epsilon' a^+[u,u] + C_{\epsilon'}\|u\|$$

By (13) and (14) we have (12) and consequently (10). Finally, (8) follows from (9) and (10), and Lemma 4.1 is proved.

The form

$$a_{(\alpha)}[u,v] + c(u,v), \quad u,v \in D(a_{(\alpha)}),$$

is positive definite for a sufficiently large c, and the above considerations show that the norms

$$\| u \|_{a_{(\alpha)}} = (a_{(\alpha)}[u, u] + c \| u \|^2)^{1/2} \quad \text{and} \quad \| u \|_{a^+}$$

are equivalent. Hence if we close $D(a_{(\alpha)})$ in the norm $\| \cdot \|_{a_{(\alpha)}}$, then a Hilbert space is obtained which depends on the submatrix

$$(\alpha)_R = \begin{pmatrix} \alpha_{0,l+1} & \cdots & \alpha_{0,2l+1} \\ \cdot & & \cdot \\ \cdot & & \cdot \\ \cdot & & \cdot \\ \alpha_{l,l+1} & & \alpha_{l,2l+1} \end{pmatrix}$$

but does not depend on the submatrix

$$(\alpha)_L = \begin{pmatrix} \alpha_{00} & \cdots & \alpha_{0l} \\ \cdot & & \cdot \\ \cdot & & \cdot \\ \cdot & & \cdot \\ \alpha_{l0} & & \alpha_{ll} \end{pmatrix}.$$

We denote this Hilbert space by $H_{(\alpha)R}$. By closure of $a_{(\alpha)}[u, v], D(a_{(\alpha)})$, we obtain a closed form which we denote by

$$\bar{a}_{(\alpha)}[u, v], \ D(\bar{a}_{(\alpha)}) = H_{(\alpha)R} \, .$$

Obviously we have

$$H_{(\alpha)R} \subseteqq W^n_{2,\mathrm{loc}}(0, \infty) \, .$$

The form $\bar{a}_{(\alpha)}[u, v]$ gives rise to a semibounded and self-adjoint operator $A_{(\alpha)}$ where, according to Theorem 1.5, the relationship

$$(A_{(\alpha)}u, v) = \bar{a}_{(\alpha)}[u, v], \quad u \in D(A_{(\alpha)}), \quad v \in H_{(\alpha)R} \, ,$$

holds. Thus different matrices (α) can define the same operator $A_{(\alpha)}$.

As regards the form of $A_{(\alpha)}$ we prove

Lemma 4.2

The functions of $D(A_{(\alpha)})$ belong to $W^{2n}_{2,\mathrm{loc}}(0, \infty)$, and $A_{(\alpha)}$ is represented by

$$A_{(\alpha)}u = \sum_{k=0}^{n} (-1)^k [a_k(x)u^{(k)}]^{(k)} \, , \quad u \in D(A_{(\alpha)}) \, . \tag{15}$$

Proof

By (3) of Theorem 1.5 every $u \in C_0^\infty(0, \infty)$ belongs to $D(A_{(\alpha)})$, so that $A_{(\alpha)}$ is a self-adjoint extension of the operator

$$A_0 u = \sum_{k=0}^{n} (-1)^k [a_k(x) u^{(k)}]^{(k)}, \quad D(A_0) = C_0^\infty(0, \infty) .$$

Since $A_0 \subseteqq A_{(\alpha)} \subseteqq A_0^*$, we have $D(A_{(\alpha)}) \subseteqq W_{2,\mathrm{loc}}^{2n}(0, \infty)$ by Theorem 2.1. Finally, (15) follows by integration by parts from

$$(A_{(\alpha)} u, v) = \bar{a}_{(\alpha)} [u, v] = \sum_{k=0}^{n} \int_0^\infty a_k(x) u^{(k)}(x) \bar{v}^{(k)}(x) \mathrm{d}x$$

$$= \left(\sum_{k=0}^{n} (-1)^k [a_k(x) u^{(k)}]^{(k)}, v \right)$$

for $u \in D(A_{(\alpha)})$ and $v \in C_0^\infty(0, \infty)$. This proves Lemma 4.2.

To describe $A_{(\alpha)}$ in terms of boundary conditions determined by the matrix (α) at the point $x = 0$, we first establish that the functions of $D(A_{(\alpha)})$ are l-times continuously differentiable from the right at $x = 0$. We have

Lemma 4.3

The restriction of $H_{(\alpha)R}$ to the basic interval $[0, x_0]$ is continuously embedded in $C^l[0, x_0], 0 < x_0 < \infty$.

Proof

Since (11) holds for the functions of $D(a_{(\alpha)})$, it holds for the functions of $H_{(\alpha)R}$ also.

Lemma 4.4

Let $u \in D(A_{(\alpha)})$, and define

$$u^{[2n-\nu]}(x) = \sum_{k=\nu}^{n} (-1)^{k-\nu} [a_k(x) u^{(k)}(x)]^{(k-\nu)}, \quad 1 \leqslant \nu \leqslant l+1 . \quad (16)$$

Then the limits

$$\lim_{x \downarrow 0} u^{[2n-\nu]}(x) = u^{[2n-\nu]}(0)$$

exist.

Proof

In the case $\nu = 1$ we have

$$u^{[2n-1]}(x) = u^{[2n-1]}(1) - \int_x^1 \left(\sum_{k=1}^{n} (-1)^{k-1} [a_k(t)u^{(k)}(t)]^{(k)} \right) \mathrm{d}t$$

$$= u^{[2n-1]}(1) + \int_x^1 \left(\sum_{k=0}^{n} (-1)^{k} [a_k(t)u^{(k)}(t)]^{(k)} \right) \mathrm{d}t - \int_x^1 a_0(t)u(t)\,\mathrm{d}t \,.$$

The integrals on the right-hand side converge as x tends to zero since, by Lemma 4.2 and (2), we have

$$\int_x^1 \left| \sum_{k=0}^{n} (-1)^{k} [a_k(t)u^{(k)}(t)]^{(k)} \right| \mathrm{d}t$$

$$\leqslant \left(\int_0^1 \left| \sum_{k=0}^{n} (-1)^{k} [a_k(t)u^{(k)}(t)]^{(k)} \right|^2 \mathrm{d}t \right)^{1/2} \leqslant \| A_{(a)}u \| < \infty$$

and

$$\int_x^1 | a_0(t)u(t) |\,\mathrm{d}t \leqslant \left(\max_{0 \leqslant x \leqslant 1} | u(x) | \right) \int_0^1 | a_0(t) |\,\mathrm{d}t$$

$$= C_u \int_0^1 | a_0(t) |\,\mathrm{d}t < \infty, \quad 0 < x < 1 \,.$$

Then the limit procedure $x \to 0$ yields

$$u^{[2n-1]}(0) = u^{[2n-1]}(1) + \int_0^1 \left(\sum_{k=1}^{n} (-1)^{k} [a_k(t)u^{(k)}(t)]^{(k)} \right) \mathrm{d}t \,.$$

Proceeding inductively we obtain the existence of $u^{[2n-\nu]}(0)$ and the formula

$$u^{[2n-\nu]}(0) = u^{[2n-\nu]}(1) + \int_0^1 \left(\sum_{k=\nu}^{n} (-1)^{k-\nu+1} [a_k(t)u^{(k)}(t)]^{(k-\nu+1)} \right) \mathrm{d}t \,,$$

$$1 \leqslant \nu \leqslant l+1 \,.$$

This proves the Lemma

Let $N_{(\alpha)_R}$ be the null-space of the operator represented by the matrix $(\alpha)_R$ in the $(l+1)$-dimensional (complex) Euclidian space and let $N_{(\alpha)_R}^{\perp}$ be the orthogonal complement of $N_{(\alpha)_R}$,

$$R^{l+1} = N_{(\alpha)_R} \oplus N_{(\alpha)_R}^{\perp} \,.$$

Further for $u \in D(A_{(\alpha)})$ let

$$Y = (Y_0, \ldots, Y_1) \in R^{l+1}$$

and

$$Y_0^*, \ldots, Y_l^*) \in R^{l+1}$$

be the vectors with the components

$$Y_j = u^{(j)}(0)$$

and

$$Y_j^* = u^{[2n-j-1]}(0) - \sum_{i=0}^{l} \alpha_{ij} u^{(i)}(0), \quad 0 \leqslant j \leqslant l\,.$$

Theorem 4.5 [42]
The functions $u(x) \in D(A_{(\alpha)})$ satisfy the boundary conditions $Y \in N_{(\alpha)R}$ and $Y^* \in N_{(a)R}^{\perp}$ at $x = 0$.

Conversely, if a function $u \in W_{2,\mathrm{loc}}^{2n}(0, \infty) \cap L_2(0, \infty)$ with continuous derivatives $u^{(i)}(x)$, $0 \leqslant i \leqslant l$, and quasi-derivatives $u^{[\nu]}(x)$, $2n - l - 1 \leqslant \nu \leqslant 2n - 1$, from the right at $x = 0$ and with the property $a[u] \in L_2(0, \infty)$ satisfies these boundary conditions, then it belongs to $D(A_{(\alpha)})$.

Proof
Let $u \in D(A_{(\alpha)})$. Since by Lemma 4.3 the embedding from $H_{(\alpha)R}$, restricted to the basic interval $[0, x_0]$, into $C^l[0, x_0]$ is continuous, the equations

$$\sum_{j=0}^{l} \alpha_{i,l+1+j}\, u^{(j)}(0) = 0, \quad 0 \leqslant i \leqslant l,$$

involved in the description of $D(a_{(\alpha)})$, are also valid for the functions of $H_{(\alpha)R}$. Hence $Y \in N_{(\alpha)R}$.

From

$$(A_{(\alpha)}u, v) = \bar{a}_{(\alpha)}[u, v], \quad u \in D(A_{(\alpha)}), \quad v \in D(a_{(\alpha)})\,,$$

we have

$$\lim_{x \downarrow 0} \sum_{k=0}^{n} \int_x^{\infty} (-1)^k \,[a_k(t) u^{(k)}(t)]^{(k)} \bar{v}(t)\,\mathrm{d}t$$

$$= \sum_{k=0}^{n} \int_0^{\infty} a_k(t) u^{(k)} \bar{v}^{(k)}\,\mathrm{d}t + \sum_{i,j=0}^{l} \alpha_{ij} u^{(i)}(0) \bar{v}^{(j)}(0)$$

and, integrating by parts, we obtain

$$\lim_{x\downarrow 0}\left\{\sum_{k=1}^{n}(-1)^{k-1}[a_k(x)u^{(k)}(x)]^{(k-1)}\bar{v}(x)+\cdots+\right.$$

$$\left.+\sum_{k=l+1}^{n}(-1)^{k-l-1}[a_k(x)u^{(k)}(x)]^{(k-l-1)}\bar{v}^{(l)}(x)\right\}$$

$$+\sum_{k=0}^{n}\int_0^\infty a_k(x)u^{(k)}\bar{v}^{(k)}\,dx$$

$$=\sum_{k=0}^{n}\int_0^\infty a_k(x)u^{(k)}\bar{v}^{(k)}\,dx+\sum_{i,j=0}^{l}\alpha_{ij}u^{(i)}(0)\bar{v}^{(j)}(0).$$

Then, from Lemma 4.4, it follows that

$$\left(u^{[2n-1]}(0)-\sum_{i=0}^{l}\alpha_{i0}u^{(i)}(0)\right)\bar{v}(0)+\cdots+$$

$$\left(u^{[2n-l-1]}(0)-\sum_{i=0}^{l}\alpha_{il}u^{(i)}(0)\right)\bar{v}^{(l)}(0)=0.$$

Since, on the other hand

$$\sum_{j=0}^{l}\alpha_{i,l+1+j}\bar{v}^{(j)}(0)=0,\quad 0\leqslant i\leqslant l,$$

we have $Y^*\in N^{\perp}_{(\alpha)R}$.

Conversely, if a function $u\in W_{2,\mathrm{loc}}^{2n}(0,\infty)\cap L_2(0,\infty)$ satisfies the conditions stated in the theorem, then it follows that for $v\in D(a_{(\alpha)})$.

$$(a[u],v)=\lim_{x\downarrow 0}\sum_{k=0}^{n}\int_x^\infty(-1)^k[a_k(t)u^{(k)}]^{(k)}\bar{v}(t)\,dt$$

$$=\lim_{x\downarrow 0}\left\{\sum_{k=1}^{n}(-1)^{k-1}[a_k(x)u^{(k)}]^{(k-1)}\bar{v}(x)+\cdots+\right.$$

$$\left.+\sum_{k=l+1}^{n}(-1)^{k-l-1}[a_k(x)u^{(k)}(x)]^{(k-l-1)}\bar{v}^{(l)}(x)+\sum_{k=0}^{n}\int_x^\infty a_k u^{(k)}\bar{v}^{(k)}\,dt\right\}$$

$$=\sum_{k=0}^{n}\int_0^\infty a_k(t)u^{(k)}\bar{v}^{(k)}\,dt+\sum_{i,j=0}^{l}\alpha_{ij}u^{(i)}(0)\bar{v}^{(j)}(0).$$

Hence

$$(a[u], v) = \bar{a}_{(\alpha)}[u, v], \quad v \in D(a_{(\alpha)}).$$

Then, since $D(a_{(\alpha)})$ is dense in $D(\bar{a}_{(\alpha)}) = H_{(\alpha)R}$, u belongs to $D(A_{(\alpha)})$ by Theorem 1.5, and we have $A_{(\alpha)}u = a[u]$. This proves Theorem 4.5.

Note 4.6

1) If $(\alpha)_R$ is the zero matrix, then $N_{(\alpha)R} = R^{l+1}$ and there is no restriction on Y. Then $N^{\perp}_{(\alpha)R} = \{0\}$ gives

$$\sum_{i=0}^{l} \alpha_{ij} u^{(i)}(0) = u^{[2n-j-1]}(0), \quad 0 \leqslant j \leqslant l, \tag{17}$$

as the boundary condition for $u \in D(A_{(\alpha)})$.

2) If $(\alpha)_R$ is a regular matrix, then $N_{(\alpha)R} = \{0\}$, and we have

$$u(0) = u'(0) = \cdots = u^{(l)}(0) = 0. \tag{18}$$

When $N^{\perp}_{(\alpha)R} = R^{l+1}$ the requirement $Y^* \in N^{\perp}_{(\alpha)R}$ includes no additional conditions.

3) In the case $l = 0, n = 1$ the symbol $a[\cdot]$ is the Sturm–Liouville differential expression

$$-\frac{d}{dx} a_1(x) \frac{d}{dx} + a_0(x), \quad 0 < x < \infty.$$

If the element α_{01} of the matrix $(\alpha) = (\alpha_{00}\ \alpha_{01})$ is equal to zero, then by (17) we have

$$\alpha_{00} u(0) = u^{[1]}(0) = \lim_{x \downarrow 0} a_1(x) u'(x).$$

If $\alpha_{01} \neq 0$, then $u(0) = 0$ by (18). Both these cases are covered by the boundary condition

$$u(0) \cos \varphi - a_1(0) u'(0) \sin \varphi = 0, \quad 0 \leqslant \varphi < \pi,$$

where the formal product $a_1(0)u'(0)$ stands for the limit $\lim_{x \downarrow 0} a_1(x)u'(x)$.

NON-EXISTENCE OF EIGENVALUES

Next we turn to the question of what influence the boundary conditions described in Theorem 4.5 have on the existence of eigenvalues of $A_{(\alpha)}$. Here we may assume that the eigenfunctions of $A_{(\alpha)}$ are real-valued.

Theorem 4.7

We impose the following conditions on the coefficients.

i) (1), (2), (3), and $a_i(x) \in C^1(0, \infty)$, $i = 0, 1$. (19)

ii) There are constants c_1, c_2, where $0 < c_1 < 1$, $1 < c_2$, and $X > 0$ such that

$$c_1 \mid a_k(x) \mid \leqslant \mid a_k(\tau x) \mid \leqslant c_2 \mid a_k(x) \mid, \quad 0 \leqslant k \leqslant n, \tag{20}$$

for all $x > X$ and τ, $1 \leqslant \tau \leqslant 2$.

iii) There are a constant $C > 0$ and a point $x_1 > 0$ such that

$$x \mid a_k'(x) \mid \leqslant C \mid a_k(x) \mid, \quad 0 < x < x_1, \quad 0 \leqslant k \leqslant n. \tag{21}$$

iv) $(\alpha)_R$ is a diagonal matrix. (22)

v) If $j^*, 0 \leqslant j^* \leqslant l$, is the highest index for which $a_{j^*}(x)$ takes negative values and $k^*, l + 1 \leqslant k \leqslant n$, is the lowest index such that

$$\liminf_{x \to \infty} a_{k^*}(x) x^{-\alpha k^*} > 0,$$

then there exists a ρ with

$$k^*(\alpha - 2) \leqslant \rho \leqslant j^*(\alpha - 2), \quad \alpha \leqslant 2, \tag{23}$$

and a real λ such that the functions

$$(a_0(x) - \lambda) x^{-\rho}, \quad a_k(x) x^{-(\rho + 2k)}, \quad 1 \leqslant k \leqslant n, \quad 0 < x < \infty, \tag{24}$$

are decreasing.† For this ρ, let the form

$$\sum_{i,j=0}^{l} (2j + 1 + \rho) \alpha_{ij} \xi_i \bar{\xi}_j, \quad \xi = (\xi_0, \ldots, \xi_l) \in N_{(\alpha)_R}, \tag{25}$$

be positive semidefinite. If no coefficient is negative, then the assumptions (3) are absent. In this case let ρ be restricted by $\rho \leqslant 0$ in (24) and (25).

†The restriction (23) for ρ is consistent with the fact that on the one hand the conditions (3) are imposed and on the other hand the monotony of the functions (24) is supposed.

If the conditions i) – v) are satisfied, then λ and numbers larger than λ are not eigenvalues of $A_{(\alpha)}$. If the above conditions are satisfied with $\rho = 0$, the spectrum of $A_{(\alpha)}$ is continuous.

Proof

The proof is given in several stages.

1) We show that in the neighbourhood of $x = 0$ the behaviour of the coefficients $a_k(x)$ can be described as follows.

There exist positive constants c_1 and c_2 independent of the parameter τ, $1 \leqslant \tau \leqslant 2$, such that

$$c_1 \, |a_k(x)| \leqslant |a_k(\tau x)| \leqslant c_2 \, |a_k(x)| \, , \quad 1 \leqslant \tau \leqslant 2,$$
$$0 < x < \tfrac{1}{2} x_1, \quad 0 \leqslant k \leqslant n \, . \tag{26}$$

An index k, $0 \leqslant k \leqslant n$, is singled out. If $a_k(x) = 0$, $x \in (0, x_1)$, then we have (26) for arbitary positive constants c_1 and c_2. If there is a point $\bar{x}$, $0 < \bar{x} < x_1$, such that $a_k(\bar{x}) > 0$, then we let (ξ_1, ξ_2), $0 \leqslant \xi_1 < \bar{x} < \xi_2 \leqslant x_1$, be the largest interval where the continuous function $a_k(x)$ is positive. Then from (21) we have

$$-Cx^{-1} \leqslant a_k'(x)/a_k(x) \leqslant Cx^{-1}, \quad x \in (\xi_1, \xi_2) \, , \tag{27}$$

and integration from $\bar{x}$ to $x \in (\xi_1, \xi_2)$ gives

$$a_k(\bar{x}) \left(\frac{x}{\bar{x}} \right)^{-C} \leqslant a_k(x) \leqslant a_k(\bar{x}) \left(\frac{x}{\bar{x}} \right)^{C}, \quad \bar{x} \leqslant x < \xi_2 \, ,$$

and

$$a_k(x) \left(\frac{x}{\bar{x}} \right)^{-C} \geqslant a_k(x) \geqslant a_k(\bar{x}) \left(\frac{x}{\bar{x}} \right)^{C}, \quad \xi_1 < x \leqslant \bar{x} \, .$$

From these estimates we obtain

$$a_k(\xi_2) \geqslant a_k(\bar{x}) \left(\frac{\xi_2}{\bar{x}} \right)^{-C}$$

and

$$a_k(\xi_1) \geqslant a_k(\bar{x}) \left(\frac{\xi_1}{\bar{x}} \right)^{C} .$$

It now follows that $\xi_1 = 0$ and $\xi_2 = x_1$. If we now choose any $\bar{x} \in (0, \frac{1}{2}x_1)$ and put $x = \tau\bar{x}$, $1 \leqslant \tau \leqslant 2$, we obtain

$$2^{-C} a_k(\bar{x}) \leqslant a_k(\tau\bar{x}) \leqslant 2^C a_k(\bar{x}), \quad 1 \leqslant \tau \leqslant 2 .$$

This proves (26), with $c_1 = 2^{-C}$ and $c_2 = 2^C$, for the index k under consideration.

If there exists a point $\bar{x}$, $0 < \bar{x} < x_1$, where $a_k(\bar{x}) < 0$, then we can argue likewise. We have now proved that on $(0, \frac{1}{2}x_1)$ the sign of each coefficient $a_k(x)$ is fixed.

2) In what follows we use the scale transformation

$$\xi = \tau x, \quad 0 < x < \infty, \quad 1 < \tau < 2 .$$

Also, we define the function $\phi_r(x) = \phi_1(x/r)$, $r > 1$, where

$$\varphi_1(x) \begin{cases} = 1, & 0 \leqslant x \leqslant 1 , \\ = 0, & x \geqslant 2 , \\ \in C^\infty [0, \infty), & 0 \leqslant \varphi_1(x) \leqslant 1 . \end{cases}$$

We now assume that λ is an eigenvalue of $A_{(\alpha)}$ with $u(x)$ as a corresponding (real-valued) eigenfunction. We consider the function $u(\tau x)\varphi_r(x)$ and show that it belongs to the Hilbert space $H_{(\alpha)_R}$. Let $\{u_\nu(x)\}_{\nu=1,2,\ldots}$, $u_\nu \in D(a_{(\alpha)})$, be a sequence which approximates the eigenfunction $u(x)$ in the norm $\|\cdot\|_{a^+}$. Since $(\alpha)_R$ is a diagonal matrix, the property $Y \in N_{(\alpha)_R}$ means that the boundary conditions have the form

$$u^{(l_\kappa)}(0) = 0, \quad \kappa = 1, \cdots, K, \quad 0 \leqslant l_\kappa \leqslant l .$$

Since the $u_\nu(x)$, $\nu = 1, 2, \ldots$, satisfy these boundary conditions, so also do $u_\nu(\tau x)\varphi_r(x)$, $\nu = 1, 2, \ldots$. These functions belong to $H_{(\alpha)_R}$ and they approximate $u(\tau x)\varphi_r(x)$ as we now show using (26) and (A.9). We have

$$\begin{aligned} &\| u_\nu(\tau x)\varphi_r(x) - u_\mu(\tau x)\varphi_r(x) \|_{a^+}^2 \\ &= \sum_{k=0}^{n} \int_0^{\frac{1}{2}x_1} a_k^+(x)[(u_\nu(\tau x)\varphi_r(x) - u_\mu(\tau x)\varphi_r(x))^{(k)}]^2 \, dx \\ &+ \sum_{k=0}^{n} \int_{\frac{1}{2}x_1}^{2r} a_k^+(x)\,[(u_\nu(\tau x)\varphi_r(x) - u_\mu(\tau x)\varphi_r(x))^{(k)}]^2 \, dx \\ &+ \int_0^{2r} [u_\nu(\tau x)\varphi_r(x) - u_\mu(\tau x)\varphi_r(x)]^2 \, dx \end{aligned} \tag{28}$$

$$\leqslant c_1^{-1} \sum_{k=0}^{n} \int_0^{\frac{1}{2}x_1} a_k^+(\tau x)[(u_\nu(\tau x)\varphi_r(x) - u_\mu(\tau x)\varphi_r(x))^{(k)}]^2 \,dx$$

$$+ C_r \| u_\nu(x) - u_\mu(x) \|^2_{W_2^n(\frac{1}{2}x_1, 4r)} + \| u_\nu(x) - u_\mu(x) \|^2_{(0,4r)}$$

$$\leqslant c_1^{-1}\, 2^{2n-1} \sum_{k=0}^{n} \int_0^{x_1} a_k^+(x)[u_\nu^{(k)}(x) - u_\mu^{(k)}(x)]^2 \,dx$$

$$+ C_r' \left(\sum_{k=0}^{n} \int_{\frac{1}{2}x_1}^{4r} a_k^+(x)[u_\nu^{(k)}(x) - u_\mu^{(k)}(x)]^2 \,dx + \| u_\nu(x) - u_\mu(x) \|^2 \right)$$

$$\leqslant C_r'' \| u_\nu(x) - u_\mu(x) \|^2_{a^+}, \quad \mu, \nu = 1, 2, \ldots .$$

Since also

$$\| u_\nu(\tau x)\varphi_r(x) - u(\tau x)\varphi_r(x) \| \to 0, \quad \nu \to \infty ,$$

$u(\tau x)\varphi_r(x)$ belongs to $H_{(\alpha)R}$. Hence, by Theorem 1.5, we have

$$(A_{(\alpha)}u(x), u(\tau x)\varphi_r(x)) = \lambda(u(x), u(\tau x)\varphi_r(x))$$

$$= \bar{a}_{(\alpha)}[u(x), \ u(\tau x)\varphi_r(x)] ,$$

which gives

$$\sum_{k=0}^{n} \int_0^{\infty} a_k(x)u^{(k)}(x)[u(\tau x)\varphi_r(x)]^{(k)} \,dx + \sum_{i,j=0}^{l} \alpha_{ij}\tau^j u^{(i)}(0)u^{(j)}(0)$$
$$= \lambda \int_0^{\infty} u(x)u(\tau x)\varphi_r(x) \,dx . \tag{29}$$

Similarly we can prove that $u(\tau^{-1} x)\varphi_r(\tau^{-1}x) \in H_{(\alpha)R}$ and

$$\sum_{k=0}^{n} \int_0^{\infty} a_k(x)u^{(k)}(x)[u(\tau^{-1}x)\varphi_r(\tau^{-1}x)]^{(k)} \,dx + \sum_{i,j=0}^{l} \alpha_{ij}\tau^{-j}u^{(i)}(0)u^{(j)}(0)$$

$$= \lambda \int_0^{\infty} u(x)u(\tau^{-1}x)\varphi_r(\tau^{-1}x) \,dx .$$

If we substitute $x = \tau\xi$ in the integrals in the last equation and after that ξ is denoted by x again, we obtain

$$\sum_{k=0}^{n} \tau^{2(n-k)} \int_0^\infty a_k(\tau x) u^{(k)}(\tau x)[u(x)\varphi_r(x)]^{(k)}\,\mathrm{d}x + \sum_{i,j=0}^{l} \alpha_{ij}\tau^{2n-j-1} u^{(i)}(0) u^{(j)}(0) \tag{30}$$

$$= \lambda\tau^{2n} \int_0^\infty u(x)u(\tau x)\varphi_r(x)\,\mathrm{d}x\,.^{\dagger}$$

We now subtract (29) from (30), multiply the difference by $\epsilon^{-1}(=(\tau-1)^{-1})$, and calculate the limit as $\epsilon \to 0$. On the right-hand side the result of this procedure is clearly

$$2n\lambda \int_0^\infty u^2(x)\varphi_r(x)\,\mathrm{d}x \tag{31}$$

and, from the sum $\sum_{i,j=0}^{l}$, we obtain

$$\sum_{i,j=0}^{l} (2n-2j-1)\alpha_{ij} u^{(i)}(0) u^{(j)}(0)\,. \tag{32}$$

To deal with the sum $\sum_{k=0}^{n}$ we first examine

$$\lim_{\epsilon\to 0} \int_0^\infty \epsilon^{-1}\left[\tau^{2(n-k)} a_k(\tau x) - a_k(x)\right] u^{(k)}(x) u^{(k)}(\tau x)\varphi_r(x)\,\mathrm{d}x\,. \tag{33}$$

Let us choose any η, where $0 < \eta < \frac{1}{2}x_1$. Since the integrand tends uniformly to

$$[2(n-k)a_k(x) + x a_k'(x)]\,(u^{(k)}(x))^2\,\varphi_r(x)$$

on $[\eta, 2r]$ as $\epsilon \to 0$, (33) is

$$\int_0^\infty [2(n-k)a_k + x a_k'](u^{(k)})^2 \varphi_r\,\mathrm{d}x \tag{34}$$

† $u^{(k)}(\tau x)$ means $\dfrac{\mathrm{d}^k}{\mathrm{d}x^k} u(\tau x)$ here.

if the integral

$$\mathscr{F}_k(\eta) = \int_0^\eta \epsilon^{-1} [\tau^{2(n-k)} a_k(\tau x) - a_k(x)] u^{(k)}(x) u^{(k)}(\tau x) \varphi_r(x) \, dx$$

tends to zero uniformly with respect to $\epsilon (0 < \epsilon < 1)$ as $\eta \to 0$. This we now prove and we use (21), (26), and the mean value theorem of differential calculus. Also in what follows, $x < x_\epsilon < x + \epsilon x$ and r is chosen sufficiently large.

$$|\mathscr{F}_k(\eta)| = \left| \int_0^\eta \epsilon^{-1} [\tau^{2(n-k)} a_k(\tau x) \mp a_k(\tau x) - a_k(x)] u^{(k)}(x) u^{(k)}(\tau x) \, dx \right|$$

$$\leqslant \int_0^\eta | \epsilon^{-1} (\tau^{2(n-k)} - 1) a_k(\tau x) u^{(k)}(x) u^{(k)}(\tau x) | \, dx$$

$$+ \int_0^\eta x_\epsilon \, | a_k'(x_\epsilon) | \, | u^{(k)}(x) u^{(k)}(\tau x) | \, dx$$

$$\leqslant C_1 \int_0^\eta | a_k(x) |^{1/2} \, | a_k(\tau x) |^{1/2} \, | u^{(k)}(x) u^{(k)}(\tau x) | \, dx$$

$$+ \int_0^\eta x_\epsilon \, | a_k'(x_\epsilon) | \, | u^{(k)}(x) u^{(k)}(\tau x) | \, dx$$

$$\leqslant C_1 \left(\int_0^\eta | a_k(x) | [u^{(k)}(x)]^2 \, dx \right)^{1/2} \left(\int_0^\eta | a_k(\tau x) | [u^{(k)}(\tau x)]^2 \, dx \right)^{1/2}$$

$$+ C_2 \int_0^\eta | a_k(x_\epsilon) | \, | u^{(k)}(x) u^{(k)}(\tau x) | \, dx$$

$$\leqslant C_3 \int_0^{2\eta} | a_k(x) | [u^{(k)}(x)]^2 \, dx \leqslant C_3 \int_0^{2\eta} a_k^+(x) [u^{(k)}]^2 \, dx + C_3 \int_0^{2\eta} | a_k^-(x) | [u^{(k)}]^2 \, dx$$

$$\leqslant C_3 \int_0^{2\eta} a_k^+(x) [u^{(k)}]^2 \, dx + C_3 \left(\max_{x \in [0, 2\eta]} [u^{(k)}]^2 \right) \int_0^{2\eta} | a_k^-(x) | \, dx . \tag{35}$$

Hence, by (11) and (12) we obtain finally

$$| \mathscr{F}_k(\eta) | \leqslant C \int_0^{2\eta} a_k^+ [u^{(k)}]^2 \, dx + C \| u \|_{a^+}^2 \int_0^{2\eta} | a_k^-(x) | \, dx = o(1)$$

uniformly in ϵ as $\eta \to 0$.

The remainder of $\sum_{k=0}^{n}$ contains terms of the form

$$S_{jk}(\epsilon, r) = \int_0^\infty \epsilon^{-1}\,[\tau^{2(n-k)} a_k(\tau x) u^{(k)}(\tau x) u^{(k)}(\tau x) u^{(k-j)}(x) -$$

$$a_k(x) u^{(k)}(x) u^{(k-j)}(\tau x)] \binom{k}{j} \varphi_r^{(j)}(x)\, \mathrm{d}x, \quad 1 \leqslant j \leqslant k \leqslant n\,.$$

Since $j \geqslant 1$ the support of the integrand here lies in $[r, 2r]$. The limit procedure $\epsilon \to 0$ (where in the integrand the De l'Hospital rule can be used) gives

$$\lim_{\epsilon\to 0} S_{jk}(\epsilon, r) = \int_0^\infty [2(n-k) a_k u^{(k)} u^{(k-j)} + x a_k' u^{(k)} u^{(k-j)} + k a_k u^{(k)} u^{(k-j)} +$$

$$x a_k u^{(k+1)} u^{(k-j)} - (k-j) a_k u^{(k)} u^{(k-j)} - x a_k u^{(k)} u^{(k-j+1)}] \binom{k}{j} \varphi_r^{(j)}\, \mathrm{d}x\,.$$

By integration by parts the derivative a_k' is removed, and the result is

$$\lim_{\epsilon\to 0} S_{jk}(\epsilon, r) = S_{jk}(0, r)$$

$$= \binom{k}{j} \int_r^{2r} ([2(n-k) + j - 1] a_k u^{(k)} u^{(k-j)} - 2x a_k u^{(k)} u^{(k-j+1)}) \varphi_r^{(j)}\, \mathrm{d}x -$$

$$\binom{k}{j} \int_r^{2r} x a_k u^{(k)} u^{(k-j)} \varphi_r^{(j+1)}\, \mathrm{d}x\,. \tag{36}$$

We aim to show that $S_{jk}(0, r) \to 0$ as $r \to \infty$. Using $|\varphi_r{}^{(j)}(x)| \leqslant C r^{-j}$ on $[r, 2r]$, the Schwarz inequality, (10) and (20), we have

$$\left| \int_r^{2r} a_k u^{(k)} u^{(k-j)} \varphi_r^{(j)}\, \mathrm{d}x \right|^2$$

$$\leqslant C_1 \left(\int_r^{2r} a_k^+ [u^{(k)}]^2\, \mathrm{d}x + \int_r^{2r} |a_k^-| [u^{(k)}]^2\, \mathrm{d}x \right) \times$$

$$\left(\int_r^{2r} x^{-2k} \,|\, a_k \,|\, x^{2(k-j)} [u^{(k-j)}]^2\, \mathrm{d}x \right) \tag{37}$$

$$\leqslant C_2 \,\|\, u \,\|_{a^+}^2 r^{-2k} \,|\, a_k(r) | \int_r^{2r} x^{2(k-j)} [u^{(k-j)}]^2\, \mathrm{d}x\,.$$

If we substitute $x = r^{-1}\xi$ in the estimate

$$\int_1^2 [u^{(k-j)}(x)]^2\,dx \leqslant C\left(\int_1^2 [u^{(k)}(x)]^2\,dx + \int_1^2 [u(x)]^2\,dx\right), \qquad 1 \leqslant j \leqslant k,$$

which follows from (A.9) and if then ξ is denoted by x and $u(\xi/r)$ by $u(x)$, we obtain

$$\int_r^{2r} x^{2(k-j)}[u^{(k-j)}(x)]^2\,dx \leqslant C\left(\int_r^{2r} x^{2k}[u^{(k)}(x)]^2\,dx + \int_r^{2r} [u(x)]^2 dx\right). \tag{38}$$

On substituting (38) in (37) and using (20), we obtain

$$\begin{aligned}
&\left|\int_r^{2r} a_k u^{(k)} u^{(k-j)} \varphi_r^{(j)}\,dx\right|^2 \\
&\leqslant C_1 \| u \|_{a^+}^2 r^{-2k} \,|\, a_k(r)|\left(\int_r^{2r} x^{2k}[u^{(k)}]^2\,dx + \int_r^{2r} u^2\,dx\right) \qquad (39)\\
&\leqslant C_2 \| u \|_{a^+}^2\left(\int_r^{2r} |\, a_k(x)|[u^{(k)}]^2\,dx + \int_r^{2r} u^2\,dx\right) \\
&\leqslant C_2 \| u \|_{a^+}^2 \| u \|_{a^+,(r,2r)}^2 + C_2 \| u \|_{a^+}^2 \int_r^{2r} |\, a_k^-(x)|[u^{(k)}]^2\,dx .
\end{aligned}$$

From (14) we have

$$\int_r^{2r} |\, a_k^-(x)|[u^{(k)}]^2\,dx \leqslant C \| u \|_{a^+,(r,\infty)}^2 , \tag{40}$$

and hence from (39)

$$\left|\int_r^{2r} a_k u^{(k)} u^{(k-j)} \varphi_r^{(j)}\,dx\right|^2 \leqslant C \| u \|_{a^+}^2 \| u \|_{a^+,(r,\infty)}^2 \to 0, \quad r \to \infty . \tag{41}$$

The other terms in (38) can be treated similarly, and so

$$S_{jk}(0, r) \to 0, \quad r \to \infty . \tag{42}$$

Summarizing the results (31), (32), (34), and (42), we have proved that

$$\sum_{k=0}^{n} \int_0^\infty [2(n-k)a_k + xa_k'] [u^{(k)}]^2 \varphi_r \, dx$$

$$+ \sum_{i,j=0}^{l} (2n-2j-1)\alpha_{ij} u^{(i)}(0) u^{(j)}(0) = 2n\lambda \int_0^\infty u^2 \varphi_r \, dx + o(1), \quad (43)$$

$$r \to \infty .$$

On the other hand from

$$\bar{a}_{(\alpha)} [u(x), u(x)\varphi_r(x)] = (A_{(\alpha)} u(x), u(x)\varphi_r(x)) = \lambda(u(x), u(x)\varphi_r(x))$$

it follows that

$$\sum_{k=0}^{n} \int_0^\infty a_k(x) [u^{(k)}(x)]^2 \varphi_r(x) \, dx + \sum_{i,j=0}^{l} \alpha_{ij} u^{(i)}(0) u^{(j)}(0)$$

$$(44)$$

$$= \lambda \int_0^\infty u^2(x) \varphi_r(x) \, dx + o(1), \; r \to \infty .$$

Multiplying (44) by $2n + \rho$, where ρ is a real parameter, and combining it with (43), we obtain finally

$$\sum_{k=0}^{n} \int_0^\infty [xa_k' - (2k+\rho)a_k] [u^{(k)}]^2 \varphi_r \, dx \; -$$

$$(45)$$

$$\sum_{i,j=0}^{l} (2j+1+\rho)\alpha_{ij} u^{(i)}(0) u^{(j)}(0) + \rho\lambda \int_0^\infty u^2 \varphi_r \, dx = o(1), \; r \to \infty .$$

In (45), instead of α_{ij}, the real parts a_{ij} of α_{ij} may be used.

3) If now the conditions

i) $xa_0' + \rho(\lambda - a_0) \leqslant 0, \quad 0 < x < \infty,$

ii) $xa_k' - (2k+\rho)a_k \leqslant 0, \quad 0 < x < \infty, \quad 1 \leqslant k \leqslant n,$ (46)

iii) $\sum_{i,j=0}^{l} (2j+1+\rho) a_{ij} \xi_i \xi_j$ positive semidefinite, $\xi = (\xi_0, \ldots, \xi_l) \in N_{(\alpha)R}$,
ξ real,

are satisfied, the left-hand side of (45) is nonpositive. We will show that $u(x) = 0$, $0 < x < \infty$, so proving the theorem.

First we consider the special case

$$xa_0' + \rho(\lambda - a_0) = 0, \quad xa_k' - (2k + \rho)a_k = 0, \quad 0 < x < \infty, \quad 1 \leqslant k \leqslant n .$$

Then there are real constants α_k such that

$$a_0(x) = \lambda + \alpha_0 x^\rho , \quad a_k(x) = \alpha_k x^{\rho+2k}, \quad 0 < x < \infty, \quad 1 \leqslant k \leqslant n ,$$

and the equation $A_{(\alpha)}u = \lambda u$ becomes an Euler differential equation

$$\sum_{\nu=0}^{2n} \tilde{\alpha}_\nu x^\nu u^{(\nu)} = 0, \quad \tilde{\alpha}_\nu = \text{const}, \quad \tilde{\alpha}_{2n} > 0, \quad 0 < x < \infty , \tag{47}$$

for the solution $u(x)$ in question. By Lemma 4.2 u belongs to $W^{2n}_{2,\mathrm{loc}}(0, \infty)$ and therefore to $C^{2n-1}(0, \infty)$. Then by (47) we have $u(x) \in C^\infty(0, \infty)$. Among the solutions of the differential equation (47) there is no nontrivial one which belongs to $L_2(0, \infty)$ (see [24]).

Now we assume that there is a point $\bar{x}, 0 < \bar{x} < \infty$, where at least one of the inequalities (46) (i) or (ii) is strict. Supposing

$$\bar{x}\, a_0'(\bar{x}) + \rho(\lambda - a_0(\bar{x})) < 0, \quad 0 < \bar{x} < \infty ,$$

then because of the continuity of $a_0'(x)$ this inequality is also true in a neighbourhood $U(\bar{x})$ of $\bar{x}$. In this neighbourhood $u(x)$ must be identically zero, as follows from (45) for sufficiently large r if we recall that $\rho \leqslant 0$ by (23). Since $u(x)$ is a solution of the differential equation

$$\sum_{k=0}^{n} (-1)^k [a_k(x)u^{(k)}]^{(k)} = \lambda u , \tag{48}$$

it must be identically zero in $(0, \infty)$ because of the uniqueness of solutions of (48).†
Therefore λ is not an eigenvalue.

Now we consider the case that

$$xa_0' + \rho(\lambda - a_0) = 0, \quad xa_k' - (2k + \rho)a_k = 0, \quad 0 < x < \infty,$$
$$1 \leqslant k < \bar{k} \leqslant n ,^{‡} \tag{49}$$

while, for $k = \bar{k}$, there is a point $\bar{x}, 0 < \bar{x} < \infty$, where the strict inequality

$$\bar{x}\, a_{\bar{k}}'(\bar{x}) - (2\bar{k} + \rho)a_{\bar{k}}(\bar{x}) < 0$$

†This well-known fact in the theory of linear differential equations is also true for "generalized linear differential equations" (see [46]).

‡Editor's note: the case $\bar{k} = 1$ can also be covered by this argument.

holds. Then from (45) it follows that $u^{(\bar{k})}(x)$ vanishes in a neighbourhood $U(\bar{x})$ of $\bar{x}, 0<\bar{x}<\infty$. Hence $u(x)$ is there identical with a polynomial $u_P(x)$ whose degree is at most $\bar{k}-1$. From (49) we have

$$a_0(x)=\lambda+\alpha_0 x^\rho,\quad a_k(x)=\alpha_k x^{\rho+2k},\quad 0<x<\infty,$$
$$1\leqslant k\leqslant \bar{k}-1. \tag{50}$$

Then, since $u^{(\bar{k})}(x)=0, x\in U(\bar{x}), u(x)$ satisfies the differential equation

$$\sum_{k=0}^{\bar{k}-1}(-1)^k\alpha_k(x^{\rho+2k}u^{(k)})^{(k)}=0,\quad x\in U(\bar{x}). \tag{51}$$

Otherwise $u(x)$ is a solution of (48), where the coefficients $a_k(x), 0\leqslant k\leqslant\bar{k}-1$, are as in (50). The polynomial $u_P(x)$ clearly satisfies (51) for all positive x and since the degree of $u_P(x)$ is less than $\bar{k}$, it follows that u_P also satisfies (48) for all positive x. Thus $u(x)$ and $u_P(x)$ are both solutions of (48) and in $U(\bar{x})$ they coincide. Hence $u(x)=u_P(x)$ for all positive x. However, among polynomials only the zero function belongs to $L_2(0,\infty)$, and so λ can not be an eigenvalue of $A_{(\alpha)}$.

The conditions (46) follow from (24) and (25). If (46) (i) is satisfied for λ_0, then since $\rho\leqslant 0$ it is also satisfied for $\lambda>\lambda_0$. This completes the proof of Theorem 4.7.

Examples of Theorem 4.7 are considered below.

Note 4.8

We indicate three variants of assumption ii) of Theorem 4.7 under which the theorem is still valid. In place of (3) and (ii) let one of the conditions (ii$_1$), (ii$_2$) or (ii$_3$) hold.

$$\text{ii}_1)\ \liminf_{x\to\infty} x^{-2n}a_n(x)>0,\quad \limsup_{x\to\infty} x^{-2k}\,|a_k(x)|<\infty,\ 0\leqslant k\leqslant n. \tag{52}$$

To prove Theorem 4.7 we have to show again that $S_{jk}(0,r)\to 0, r\to\infty$. For this, we obtain (37) under the assumption (52) as follows.

$$\left|\int_r^{2r} a_k u^{(k)} u^{(k-j)}\varphi^{(j)}\,\mathrm{d}x\right|^2$$
$$\leqslant C_1\left(\int_r^{2r}|a_k|[u^{(k)}]^2\,\mathrm{d}x\right)\left(\int_r^{2r}|a_k|\,x^{-2j}[u^{(k-j)}]^2\,\mathrm{d}x\right)$$
$$\leqslant C_2\left(\int_r^{2r}x^{2k}[u^{(k)}]^2\,\mathrm{d}x\right)\left(\int_r^{2r}x^{2(k-j)}[u^{(k-j)}]^2\,\mathrm{d}x\right)$$

$$\leqslant C_3\left(\int_r^{2r} x^{2n}[u^{(n)}]^2\,dx + \int_r^{2r} u^2\,dx\right)^2$$

$$\leqslant C_4\left(\int_r^{2r} a_n(x)[u^{(n)}]^2\,dx + \int_r^{2r} u^2\,dx\right)^2 \leqslant C_4 \,\|\, u \,\|^4_{a^+,(r,2r)} = o(1),$$

$$r \to \infty\,.$$

ii$_2$) $\liminf\limits_{x\to\infty} a_n(x) > 0$, $\liminf\limits_{x\to\infty} a_k(x) > -\infty$, $0 \leqslant k \leqslant n-1$,

$$\limsup_{x\to\infty} x^{-2} a_k(x) < \infty,\ 1 \leqslant k \leqslant n\,. \tag{53}$$

With these assumptions and since $1 \leqslant j \leqslant k \leqslant n$ the estimate (37) is obtained as follows.

$$\left|\int_r^{2r} a_k u^{(k)} u^{(k-j)} \varphi^{(j)}\,dx\right|^2$$

$$\leqslant C_1\left(\int_r^{2r} |\,a_k\,|[u^{(k)}]^2\,dx\right)\left(\int_r^{2r} |\,a_k|\,x^{-2j}[u^{(k-j)}]^2\,dx\right)$$

$$\leqslant C_2\left(\int_r^{2r} (a_k^+ + 1)[u^{(k)}]^2\,dx\right)\left(\int_r^{2r} [u^{(k-j)}]^2\,dx\right)$$

$$\leqslant C_3\left(\int_r^{\infty} (a_k^+[u^{(k)}]^2 + [u^{(n)}]^2 + u^2)\,dx\right)\left(\int_r^{\infty} ([u^{(n)}]^2 + u^2)\,dx\right)$$

$$\leqslant C_4 \,\|\, u \,\|^4_{a^+,(r,\infty)} = o(1),\quad r\to\infty\,.$$

ii$_3$) We write

$$a_k(x) = x^{2k} f_k(x),\quad 0 < x < \infty\,.$$

where

$$f_k(x) \leqslant C f_j(x),\quad 0 < x < \infty\,,\quad 0 \leqslant j \leqslant k \leqslant n\,,\quad f_n(x) > 0\,.$$

Let us consider the case $\rho = 0$. Now we have

$$\left|\int_r^{2r} a_k u^{(k)} u^{(k-j)} \varphi^{(j)}\,dx\right|^2$$

$$\leqslant C_1\left(\int_r^{2r} a_k [u^{(k)}]^2\,dx\right)\left(\int_r^{2r} f_k x^{2(k-j)}[u^{(k-j)}]^2\,dx\right)$$

$$\leqslant C_2\left(\int_r^{2r} a_k [u^{(k)}]^2 \, dx\right)\left(\int_r^{2r} f_{k-j} x^{2(k-j)} [u^{(k-j)}]^2 \, dx\right)$$

$$\leqslant C_2\left(\int_0^{2r} a_k [u^{(k)}]^2 \, dx\right)\left(\int_r^{2r} a_{k-j} [u^{(k-j)}]^2 \, dx\right) = o(1)\,, \quad r \to \infty\,.$$

If all the functions $f_k(x)$ are decreasing and the other conditions of Theorem 4.7 are satisfied, then the spectrum of $A_{(\alpha)}$ is continuous.

Theorem 4.7 leaves open the question of where on the λ- axis the essential spectrum of $A_{(\alpha)}$ is situated. However, by combining Theorems 2.9 and 4.7 we obtain a definite result in the next theorem.

Theorem 4.9
Let the following conditions be satisfied.

i) $a_k(x) \in W_2^k(0, X)$ for every $X > 0$, $0 \leqslant k \leqslant n$, $a_i(x) \in C^1[0, \infty)$, $i = 0, 1$.

$$\inf_{0<x<\infty} a_n(x) > 0, \quad \lim_{x\to\infty} a_k(x) = b_k, \quad 0 \leqslant k \leqslant n\,.$$

ii) $(\alpha)_R$ is a diagonal matrix.

iii) If j^*, $0 \leqslant j^* \leqslant n-1$, is the largest index such that $a_{j^*}(x)$ takes negative values and k^*, $j^* < k^* \leqslant n$, is the smallest index such that b_{k^*} is positive, then let a value ρ, $-2k^* \leqslant \rho \leqslant -2j^*$,[†] and a real λ^* exist such that the functions

$$(a_0(x) - \lambda^*)x^{-\rho} \quad \text{and} \quad a_k(x)x^{-(\rho+2k)}, \quad 1 \leqslant k \leqslant n\,, \tag{54}$$

are decreasing. For this ρ let the form

$$\sum_{i,j=0}^{l} (2j + 1 + \rho) a_{ij} \xi_i \bar{\xi}_j, \quad \xi = (\xi_0\,, \ldots, \xi_l) \in N_{(\alpha)R}\,,$$

be positive semidefinite.

Then the essential spectrum of $A_{(\alpha)}$ is the interval

$$[\Lambda, \infty), \quad \Lambda = \inf_{0<t<\infty} \sum_{k=0}^{n} b_k t^{2k}\,,$$

and on the interval $[\lambda^*, \infty)$ there is no eigenvalue of $A_{(\alpha)}$.

† See the footnote on page 139.

Proof

$A_{(\alpha)}$ is a self-adjoint extension of the operator A_0 considered in Theorem 2.9 with $a_k(x) \equiv 0, n+1 \leqslant k \leqslant 2n-1$. By this theorem we have

$$\sigma_e(A_{(\alpha)}) = [\Lambda, \infty) .$$

We can also check that conditions (i) (with $l+1=n$ and $\alpha = 0$ in (3)), (iv), and (v) of Theorem 4.7 are satisfied. Because of the boundedness of the coefficients we can now dispense with conditions (ii) and (iii) of Theorem 4.7. This follows from the estimates (28), (35), and (37) if we use

$$c_1 \| \cdot \|_{a^+} \leqslant \| \cdot \|_{W_2^n(0,\infty)} \leqslant c_2 \| \cdot \|_{a^+} , \quad c_1, c_2 > 0 ,$$

in the present case. Hence we can apply Theorem 4.7 and with that we get the second statement of Theorem 4.9.

If $\rho = 0$, then $A_{(\alpha)}$ possesses no eigenvalue. In the case where $\rho < 0$ it follows from the condition that the function $(a_0(x) - \lambda^*)x^{-\rho}$, $0 < x < \infty$, is decreasing that

$$\lambda^* \geqslant b_0 \geqslant \Lambda .$$

This proves Theorem 4.9.

We consider some examples.

Examples 4.10

1) $$a[\cdot] = \sum_{k=0}^{3} (-1)^k \frac{d^k}{dx^k} a_k(x) \frac{d^k}{dx^k} , \quad 0 < x < \infty ,$$

$$l = 1, \quad (\alpha) = \begin{pmatrix} \alpha_{00} & \alpha_{01} & 0 & 0 \\ \alpha_{10} & \alpha_{11} & 0 & 1 \end{pmatrix} ,$$

$$\alpha_{00}, \alpha_{11} \text{ real}, \quad \alpha_{01} = \bar{\alpha}_{10} .$$

Because of the choice of $(\alpha)_R$ we have $u'(0) = 0$ for the functions $u(x) \in D(a_{(\alpha)})$. The sesquilinear form is

$$a_{(\alpha)}[u, v] = \sum_{k=0}^{3} \int_0^\infty a_k(x) u^{(k)}(x) \bar{v}^{(k)}(x) \, dx + \alpha_{00} u(0) \bar{v}(0) ,$$

$D(a_{(\alpha)}) = \{u \mid u \in C^3[0, \infty),\ u(x) = 0,\ x > X_u,\ u''(x) = 0,\ 0 \leqslant x < x_u,\ u'(0) =$

Let

$$a_k(x) = x^{2k-1} f_k(x),\ \ f_k(x) \in C^3(0,\infty),\ \ 0 \leqslant k \leqslant 3\,,$$

where the factors $f_k(x)$, $0 \leqslant k \leqslant 3$, are decreasing on $(0, \infty)$. As further conditions on the $f_k(x)$ let there be numbers $\bar{x}$ and $\bar{X}$, $0 < \bar{x} < \bar{X} < \infty$, such that the following holds.

$$f_0(x) = a_0 x^{\alpha_0},\ \ a_0 \leqslant 0,\ \ \alpha_0 > 0,\ \ 0 < x \leqslant \bar{x}\,,$$

$$f_0(x) = A_0 x^{\beta_0},\ \ A_0 < 0,\ \ 0 \leqslant \beta_0 \leqslant 1,\ \ \bar{X} \leqslant x < \infty\,,$$

$$f_1(x) = a_1 x^{\alpha_1},\ \ a_1 \leqslant 0,\ \ \alpha_1 \geqslant 0\ \text{(or } a_1 > 0, -2 < \alpha_1 < 0),\ \ 0 \leqslant x < \bar{x}\,,$$

$$f_1(x) = A_1 < 0,\ \ \bar{X} \leqslant x < \infty\,,$$

$$f_2(x) = a_2 x^{\alpha_2},\ \ \alpha_2 < -2,\ \ a_2 > 0,\ \ 0 < x \leqslant \bar{x}\,,$$

$$f_2(x) = A_2 x^{\beta_2},\ \ -1 \leqslant \beta_2 \leqslant 0,\ \ A_2 > 0,\ \ \bar{X} \leqslant x < \infty\,,$$

$$f_3(x) = a_3 x^{\alpha_3},\ \ \alpha_3 \leqslant 0,\ \ a_3 > 0,\ \ 0 < x \leqslant \bar{x}\,,$$

$$f_3(x) = A_3 x^{\beta_3},\ \ \beta_3 \leqslant 0,\ \ A_3 > 0,\ \ \bar{X} \leqslant x < \infty\,,$$

$$a_k \bar{x}^{\alpha_k} > A_k \bar{X}^{\beta_k},\ \ 0 \leqslant k \leqslant 3,\ \ \beta_1 = 0\,.$$

Then the conditions (1) and (2) imposed in Theorem 4.7 are satisfied. With $\alpha = 1$ and $k^* = 2$ condition (3) is also satisfied. The same is true of (20) and (21) because in a neighbourhood to the right of zero and for large x the coefficients are power functions. $(\alpha)_R$ is a diagonal matrix and, in (23), we have $k^* = 2$ and $j^* = 1$. Since

$$a_k(x) x^{1-2k} = f_k(x),\ \ 0 \leqslant k \leqslant n\,,$$

(24) is satisfied with $\rho = -1$ and $\lambda = 0$. We have

$$N_{(\alpha)R} = \left\{ \begin{pmatrix} \xi_0 \\ 0 \end{pmatrix} \right\},\ \ \xi_0 \text{ complex}\,,$$

so that the form (25) is positive semidefinite in a trivial manner. By Theorem 4.5 it follows that the functions $u(x) \in D(A_{(\alpha)})$ satisfy the boundary conditions

$$u'(0) = 0,\ \ u^{[5]}(0) - \alpha_{00} u(0) = 0\,.$$

The application of Theorem 4.7 gives the result that there is no nonnegative eigenvalue of $A_{(\alpha)}$. $A_{(\alpha)}$ does not depend on the elements α_{01}, α_{10}, and α_{11} of the matrix (α).

2) Let

$$A_0 u = \sum_{k=0}^{n} (-1)^k a_k [(d_k + x)^{\alpha_k} u^{(k)}]^{(k)}, \quad D(A_0) = C_0^\infty(0, \infty),$$

$$a_k = \text{const} \geqslant 0, \quad d_k > 0, \quad \alpha_k \leqslant 2k, \quad 0 \leqslant k \leqslant n, \quad a_n > 0.$$

By Theorem 4.7 with $\rho = 0$ we obtain the result that the Friedrichs extension $\hat{A}$ of A_0 possesses no eigenvalue as we now show. We put $l = n-1$. The conditions (19), (20, and (21) are satisfied. If $(\alpha)_R$ is the unit matrix we have (18). Hence the sesquilinear form, which defines the operator $A_{(\alpha)}$, is

$$a_{(\alpha)}[u, v] = \sum_{k=0}^{n} \int_0^\infty a_k (d_k + x)^{\alpha_k} u^{(k)}(x) \bar{v}^{(k)}(x) \mathrm{d}x$$

with

$$D(a_{(\alpha)}) = \{u \mid u \in C^n[0, \infty); \ u(x) = 0, \ x > X_u; \ u^{(n)}(x) = 0, \ 0 \leqslant x < x_u;$$
$$u(0) = u'(0) = \ldots = u^{(n-1)}(0) = 0\}.$$

Obviously $A_{(\alpha)}$ is the Friedrichs extension $\hat{A}$ of A_0. Since $N_{(\alpha)R} = \{0\}$ it is trivial that the form (25) is positive semidefinite. Since the functions

$$a_k (d_k + x)^{\alpha_k} x^{-2k}, \quad 0 < x < \infty, \quad 0 \leqslant k \leqslant n,$$

are decreasing, $A_{(\alpha)} = \hat{A}$ possesses no eigenvalue. If, however, there is an index r, $0 \leqslant r < n$, such that $a_r > 0$, $\alpha_r > 2r$, then the spectrum of $\hat{A}$ is discrete. This follows from Theorem 3.9 if we take $s = n$ and $\alpha = 2$ there. Then we have $\tau = 1$ and the function $a_r(x) = a_r(d_r + x)^{\alpha_r}$ satisfies (3.23). The spectrum is also discrete in the case where $\alpha_n > 2n$. This can easily be seen from the Courant variational principle (Theorem 2.6).

3)
$$a[\cdot] = \frac{\mathrm{d}^2}{\mathrm{d}x^2} a_2(x) \frac{\mathrm{d}^2}{\mathrm{d}x^2} - \frac{\mathrm{d}}{\mathrm{d}x} a_1(x) \frac{\mathrm{d}}{\mathrm{d}x} + a_0(x), \quad 0 < x < \infty,$$

$$a_2(x) \in C^2[0, \infty), \quad a_1(x), \quad a_0(x) \in C^1[0, \infty),$$

$$a_2(x) > 0, \quad a_1(x) \geqslant 0, \quad a_0(x) \geqslant 0, \quad a_k'(x) \leqslant 0, \quad 0 \leqslant x < \infty, \quad k = 0, 1, 2.$$

$$\lim_{x\to\infty} a_2(x) > 0, \quad \lim_{x\to\infty} a_0(x) = 0\,,$$

$$(\alpha) = \begin{pmatrix} \alpha_{00} & 0 & 0 & 0 \\ 0 & \alpha_{11} & 0 & 0 \end{pmatrix}, \qquad \alpha_{00},\ \alpha_{11} \text{ real}, \quad N^{\perp}_{(\alpha)_R} = \{0\}\,.$$

Then conditions (i) and (ii) of Theorem 4.9 are satisfied with $n = 2$. Since $(\alpha)_R$ is the zero matrix, the boundary conditions are given by (17). Then by (16) we obtain the boundary conditions

$$\alpha_{00}u(0) - a_1(0)u'(0) + a_2'(0)u''(0) + a_2(0)u^{(3)}(0) = 0\,,$$

$$\alpha_{11}u'(0) - a_2(0)u''(0) = 0$$

for our example. If we choose $\alpha_{00} \geqslant 0$ and $\alpha_{11} \geqslant 0$ the quadratic form

$$\alpha_{00}\xi_0^2 + 3\alpha_{11}\xi_1^2, \quad \xi_0,\ \xi_1 \text{ real}\,,$$

is positive semidefinite, and then all conditions of Theorem 4.9 are satisfied with $\rho = \lambda = 0$. The spectrum of $A_{(\alpha)}$ possesses no eigenvalue and coincides with the interval $[0, \infty)$.

In the case where

$$a_0(x) \equiv 0, \quad a_1(x) \equiv 0, \quad a_2(x) \equiv 1, \quad \alpha_{00} = \sigma^3, \quad \alpha_{11} = -\sigma, \quad \sigma > 0\,,$$

(54) is satisfied if

$$-4 \leqslant \rho \leqslant 0, \quad \lambda^* \geqslant 0\,.$$

For no ρ, however, is the form

$$(1+\rho)\sigma^3\xi_0^2 - (3+\rho)\sigma\xi_1^2$$

positive semidefinite and therefore we cannot apply Theorem 4.9. In fact there is the positive eigenvalue $\lambda = \sigma^4$ in the interior of the essential spectrum $[0, \infty)$; $u = e^{-\sigma x}$ is the corresponding eigenfunction.

THE FRIEDRICHS EXTENSION

In this chapter we have hitherto made the assumptions

$$\int_0^\delta a_{l+1}^{-1}(x)\,dx < \infty, \quad \int_0^1 |a_j(x)|\,dx < \infty, \quad 0 \leqslant j \leqslant l < n\,,$$

and, consequently, boundary conditions at $x = 0$ have arisen. In this section we abandon these assumptions. We now consider the special sesquilinear form

$$a\,[u, v] = \sum_{k=0}^{n} \int_0^\infty a_k(x) u^{(k)}(x) \bar{v}^{(k)}(x) \mathrm{d}x\,, \quad D(a) = C_0^n(0, \infty)\,, \tag{55}$$

where (1) must again be satisfied. In place of (3) we assume that

$$\begin{aligned} &\liminf_{x\to\infty} a_n(x) x^{-2n} > 0, \quad \liminf_{x\to\infty} a_j(x) x^{-2j} > -\infty\,, \\ &\liminf_{x\to 0} a_n(x) x^{-2n} > 0, \quad \liminf_{x\to 0} a_j(x) x^{-2j} > -\infty, \\ &0 \leqslant j < n\,. \end{aligned} \tag{56}$$

That the form (55) is semibounded from below and closable can be proved as in Lemma 4.1 with the following estimates.

$$\begin{aligned} \int_0^\infty | a_j^-(x) \| u^{(j)} |^2 \,\mathrm{d}x &\leqslant C \int_0^\infty x^{2j} \,| u^{(j)} |^2 \,\mathrm{d}x \\ &\leqslant C\epsilon \int_0^\infty x^{2n} \,| u^{(n)} |^2 \,\mathrm{d}x + CC_\epsilon \int_0^\infty | u |^2 \,\mathrm{d}x \\ &\leqslant \epsilon' \int_0^\infty a_n(x) \,| u^{(n)} |^2 \,\mathrm{d}x + C_{\epsilon'} \int_0^\infty | u |^2 \,\mathrm{d}x, \quad 0 < j < n\,, \end{aligned}$$

and

$$\begin{aligned} \int_0^\infty | a_0^-(x) \| u |^2 \,\mathrm{d}x &\leqslant C_1 \left(\int_0^{x_1} | u |^2 \,\mathrm{d}x + \int_{x_2}^\infty | u |^2 \,\mathrm{d}x \right) + \\ &\quad + \left(\max_{x_1 \leqslant x \leqslant x_2} | u(x) |^2 \right) \int_{x_1}^{x_2} | a_0^-(x) | \,\mathrm{d}x \\ &\leqslant C_1 \| u \|^2 + C_2 (\epsilon \| u^{(n)} \|^2_{(x_1, x_2)} + C_\epsilon \| u \|^2_{(x_1, x_2)}) \\ &\leqslant \epsilon' \int_0^\infty a_n(x) | u^{(n)} |^2 \,\mathrm{d}x + C_{\epsilon'} \| u \|^2\,, \end{aligned}$$

where $\quad | a_0^-(x) | \leqslant C_1, \quad x \in (0, x_1) \cup (x_2, \infty), \quad x_1 < x_2\,.$

The operator $\hat{A}$ defined by the closed form $\bar{a}\,[u, v]$ is the Friedrichs extension of

$$A_0 u = \sum_{k=0}^{n} (-1)^k (a_k(x) u^{(k)})^{(k)}, \quad D(A_0) = C_0^\infty(0, \infty),$$

and we have the following theorem concerning the spectrum of $\hat{A}$.

Theorem 4.11

Let the following conditions be satisfied.

i) (1) and $a_i(x) \in C^1(0, \infty)$, $i = 0, \; 1$.

ii) The functions

$$f_k(x) = a_k(x) x^{-2k}, \quad 0 < x < \infty, \quad 0 \leqslant k \leqslant n,$$

are decreasing with

$$\lim_{x\to\infty} f_n(x) = \beta_n > 0, \quad \lim_{x\to\infty} f_k(x) = \beta_k > -\infty, \quad 0 \leqslant k \leqslant n-1,$$

$$\lim_{x\to 0} f_k(x) < \infty, \quad 0 \leqslant k \leqslant n. \tag{57}$$

Then the spectrum of $\hat{A}$ is continuous and coincides with the interval

$$[\Lambda, \infty), \quad \Lambda = \inf_{0<t<\infty} \sum_{k=0}^{n} \beta_k \prod_{j=1}^{k} [t^2 + (j - \tfrac{1}{2})^2].$$

Proof

In place of $\varphi_r(x)$ as used in the proof of Theorem 4.7 now we use the auxiliary function

$$\psi_r(x) = [1 - \varphi_1(rx)]\varphi_1\left(\frac{x}{r}\right) \in C_0^\infty(0, \infty), \quad r > 0.$$

If $u(x)$ is a (real-valued) eigenfunction of $\hat{A}$, then as in the proof of Theorem 4.7 the functions

$$u(\tau x)\psi_r(x), \quad u(\tau^{-1}x)\psi_r(\tau^{-1}x)$$

are considered and the equations

$$\begin{aligned}(\hat{A}u(x), \; u(\tau x)\psi_r(x)) &= \lambda(u(x), \; u(\tau x)\psi_r(x)) \\ &= \bar{a}\,[u(x), \; u(\tau x)\psi_r(x)]\end{aligned}$$

and

$$(\hat{A}u(x),\quad u(\tau^{-1}x)\psi_r(\tau^{-1}x)) = \lambda(u(x),\quad u(\tau^{-1}x)\psi_r(\tau^{-1}x))$$
$$= \bar{a}\,[u(x),\quad u(\tau^{-1}x)\psi_r(\tau^{-1}x)]$$

form the basis of the proof. Then the proof goes similarly to that of Theorem 4.7. To get the analogous equation to (43), in place of (41), we have to prove that

$$\int_{r^{-1}}^{2r^{-1}} a_k u^{(k)} u^{(k-j)} \psi_r^{(j)}\,\mathrm{d}x \to 0,\quad r \to \infty, \tag{58}$$

and

$$\int_{r}^{2r} a_k u^{(k)} u^{(k-j)} \psi_r^{(j)}\,\mathrm{d}x \to 0,\quad r \to \infty. \tag{59}$$

We prove (58). using (57), we have

$$\left|\int_{r^{-1}}^{2r^{-1}} a_k u^{(k)} u^{(k-j)} \psi_r^{(j)}\,\mathrm{d}x\right|^2$$
$$\leqslant C_1 \left(\int_{r^{-1}}^{2r^{-1}} |a_k|\,(u^{(k)})^2\,\mathrm{d}x\right)\left(\int_{r^{-1}}^{2r^{-1}} |a_k|\,x^{-2j}(u^{(k-j)})^2\,\mathrm{d}x\right)$$
$$\leqslant C_2 \left(\int_{r^{-1}}^{2r^{-1}} x^{2k}(u^{(k)})^2\,\mathrm{d}x\right)\left(\int_{r^{-1}}^{2r^{-1}} x^{2(k-j)}(u^{(k-j)})^2\,\mathrm{d}x\right)^{\dagger}$$
$$\leqslant C_3 \left(\int_{r^{-1}}^{2r^{-1}} x^{2n}(u^{(n)})^2\,\mathrm{d}x + \int_{r^{-1}}^{2r^{-1}} u^2\,\mathrm{d}x\right)^2$$
$$\leqslant C_4 \left(\int_{r^{-1}}^{2r^{-1}} a_n(u^{(n)})^2\,\mathrm{d}x + \int_{r^{-1}}^{2r^{-1}} u^2\,\mathrm{d}x\right)^2 = o(1),\quad r \to \infty.$$

Similarly we obtain (59). Hence the analogous relation to (45),

$$\sum_{k=0}^{n} \int_0^\infty [x a_k' - (2k+\rho)a_k]\,(u^{(k)})^2\,\psi_r\,\mathrm{d}x + \rho\lambda \int_0^\infty u^2\,\psi_r\,\mathrm{d}x = o(1),\quad r \to \infty,$$

can be considered as proved. With $\rho = 0$ we have

$$\sum_{k=0}^{n} \int_0^\infty [x a_k' - 2k a_k]\,(u^{(k)})^2\,\psi_r\,\mathrm{d}x = o(1),\quad r \to \infty, \tag{60}$$

† Substitute $\xi = rx$ and use (A.9)

and from this we obtain the continuity of the spectrum of $\hat{A}$ as in the proof of Theorem 4.7.

To prove that $\sigma(\hat{A})$ is the interval in question we decompose the operator at a point $a > 0$ and apply Theorem 2.22 with

$$a_k(x) = 0, \quad a < x < \infty, \quad n + 1 \leqslant k \leqslant n^*,$$

and Theorem 4.11 is proved.

From (60) we immediately obtain the next theorem.

Theorem 4.12

Let the following conditions be satisfied.

i) (1) and $a_i(x) \in C^1(0, \infty)$, $i = 0, 1$.

ii) The functions

$$w_k(x) = a_k(x) x^{-2k}, \quad 0 < x < \infty, \quad 0 \leqslant k \leqslant n,$$

are increasing with

$$\lim_{x \to 0} w_n(x) = \beta_n > 0, \quad \lim_{x \to 0} w_k(x) = \beta_k > -\infty, \quad 0 \leqslant k \leqslant n - 1,$$

$$\lim_{x \to \infty} w_k(x) < \infty, \quad 0 \leqslant k \leqslant n.$$

Then the spectrum of the Friedrichs extension $\hat{A}$ of

$$A_0 u = a[u], \quad D(A_0) = C_0^\infty(0, \infty),$$

is continuous and coincides with the interval

$$[\Lambda, \infty), \quad \Lambda = \inf_{0 < t < \infty} \sum_{k=0}^{n} \beta_k \prod_{j=1}^{k} [t^2 + (j - \tfrac{1}{2})^2].$$

Proof

The second part of the theorem again follows by the method of decomposition of operators.

Note 4.13

We give the following further result without proof (see [39]).

Let

i) $a_k(x) \in W^k_{2,\mathrm{loc}}(0, \infty)$, $2 \leqslant k \leqslant n$, $a_i(x) \in C^1(0, \infty)$, $i = 0, 1$,
$a_k(x) \leqslant 0$, $0 \leqslant k \leqslant n-1$, $a_n(x) > 0$, $0 < x < \infty$.

ii) There exist a constant $C > 0$ and values x_0 and X_0, $0 < x_0 < X_0 < \infty$, such that

$$x\,|\,a_k'(x)| \leqslant C a_k(x), \quad x \in (0, x_0) \cup (X_0, \infty), \quad 0 \leqslant k \leqslant n.$$

If the functions $x^{-2k} a_k(x)$ all are either decreasing or increasing on $(0, \infty)$, then the spectrum of the Friedrichs extension $\hat{A}$ of

$$A_0 u = \sum_{k=0}^{n} (-1)^k (a_k(x) u^{(k)})^{(k)}, \quad D(A_0) = C_0^\infty(0, \infty),$$

is continuous.

Note 4.14
If we consider differential operators whose basic interval is the whole axis and aim to formulate similar theorems on the continuity of the spectrum, then in place of the transformation $\xi = \tau x$, which we have used hitherto and which has been the fundamental instrument in obtaining (45), the transformation $\xi = x + h$, $h > 0$, must now be used (see [39]). Then boundary conditions do not appear. In this way we can obtain the following result for instance.

Let

i) $a_k(x) \in W^k_{2,\mathrm{loc}}(-\infty, \infty)$, $2 \leqslant k \leqslant n$, $a_n(x) > 0$, $-\infty, < x < \infty$,
$a_i(x) \in C^1(-\infty, \infty)$, $i = 0, 1$,

ii) $\limsup_{|x| \to \infty} x^{-4} a_k(x) < \infty$, $1 \leqslant k \leqslant n$,

iii) $a_k'(x) \leqslant 0$, $0 \leqslant k \leqslant n$ (or $a_k'(x) \geqslant 0$, $0 \leqslant k \leqslant n$), $-\infty < x < \infty$,

with

$$\lim_{x \to \infty} a_k(x) = b_k > -\infty, \quad 0 \leqslant k \leqslant n, \quad b_n > 0,$$

$$\left(\lim_{x \to -\infty} a_k(x) = b_k > -\infty, \quad 0 \leqslant k \leqslant n, \quad b_n > 0 \right)$$

respectively. The the spectrum of the Friedrichs extension of

$$A_0 u = \sum_{k=0}^{n} (-1)^k (a_k(x) u^{(k)})^{(k)}, \quad D(A_0) = C_0^\infty(-\infty, \infty),$$

is continuous and coincides with the interval

$$[\Lambda, \infty), \quad \Lambda = \inf_{0<t<\infty} \sum_{k=0}^{n} b_k t^{2k}.$$

Chapter 5

Sturm-Liouville Operators

This chapter is a continuation of the previous one and deals with the special case $n = 1$ in detail. For this case, results from the last chapter on the non-existence of eigenvalues are sharpened and further results in this circle of ideas are obtained. In the case $n = 1$ the differential expression $a[\cdot]$ is now written as

$$a[\cdot] = -\frac{d}{dx} p(x) \frac{d}{dx} + q(x), \quad 0 < x < \infty, \tag{1}$$

where we always assume that

$$p(x) > 0, \quad q(x) \text{ is real-valued}, \quad p(x) \in C^1(0, \infty), \quad q(x) \in L_{2,\text{loc}}(0, \infty). \tag{2}$$

The operator

$$A_0 u = a[u], \quad D(A_0) = C_0^\infty(0, \infty),$$

which, in contrast to the previous chapters, may not now be semibounded from below, always possesses at least one self-adjoint extension because the deficiency indices of A_0 are equal. In this chapter a well-known oscillation theorem [15] and the Sturm Separation Theorem are used in the proofs.

OSCILLATION AND THE SPECTRUM

Preparatory to the next chapter, where oscillation properties of differential equations will be studied, we begin by proving the oscillation theorem for differential equations of order $2n$. Accordingly, in the following definition and theorem, $a[\cdot]$ is the differential expression

$$a[\cdot] = \sum_{k=0}^{n} (-1)^k \frac{d^k}{dx^k} a_k(x) \frac{d^k}{dx^k}, \quad a_n(x) > 0, \quad 0 < x < \infty,$$

where

$$a_k(x) \in W^k_{2,\mathrm{loc}}(0, \infty) \text{ and real-valued, } k = 0, \ldots, n\,.$$

Definition 5.1
The differential equation

$$a[u] = \lambda u, \quad 0 < x < \infty, \tag{3}$$

is said to be oscillatory at ∞ if for every $X > 0$ there exist an interval $[\alpha, \beta]$, $X < \alpha < \beta < \infty$, and a corresponding non-trivial solution

$$u(x) \in \overset{\circ}{W}{}^n_2(\alpha, \beta) \cap W^{2n}_{2,\mathrm{loc}}(\alpha, \beta)$$

of (3).

Similarly the differential equation is said to be oscillatory at $x = 0$ if for every $X > 0$ there exists an interval $[\alpha, \beta]$, $0 < \alpha < \beta < X$, of the above kind.

In the contrary cases the differential equation is said to be non-oscillatory at ∞ or 0 respectively. A non-trivial solution of (3) is called oscillatory at ∞ or 0 if for every $X > 0$ there exists a zero of the solution in (X, ∞) or $(0, X)$ respectively.

There is a close connection between the nature of the spectrum on the one hand and the oscillatory nature of the differential equation (3) on the other hand. Thus we have the following theorem, the proof of which is basically taken from the account in [15, §12].

Theorem 5.2
The differential equation $a[u] = \lambda u$ is oscillatory at ∞ if and only if to the left of λ there are an infinity of points in the spectrum of any self-adjoint extension of

$$A_{(1,\infty),0}u = a[u], \quad D(A_{(1,\infty),0}) = C^\infty_0(1, \infty)\,.$$

The differential equation is oscillatory at 0 if and only if the same is true for any self-adjoint extension of $A_{(0,1),0}$,

$$A_{(0,1),0}u = a[u], \quad D(A_{(0,1),0}) = C^\infty_0(0, 1)\,.$$

Proof
We prove the theorem for $x = \infty$. Without loss of generality, we let $\lambda = 0$. First we assume that the equation $a[u] = 0$ is oscillatory at ∞. For an arbitrary $X > 0$ let

$$u(x) \in \overset{\circ}{W}{}^n_2(\alpha, \beta) \cap W^{2n}_{2,\mathrm{loc}}(\alpha, \beta), \quad X < \alpha < \beta < \infty,$$

be a non-trivial solution of $a[u] = 0$. Then $u(x)$ belongs to the energy space $\overset{\circ}{W}{}_2^n(\alpha, \beta)$ of the operator $A_{(\alpha,\beta),0}$,

$$A_{(\alpha,\beta),0}u = a[u], \quad D(A_{(\alpha,\beta),0}) = C_0^\infty(\alpha, \beta),$$

as well as to the domain of the adjoint operator $A^*_{(\alpha,\beta),0}$ (see Theorem 2.1). Hence $u(x)$ belongs to the domain of the Friedrichs extension $\hat{A}_{(\alpha,\beta)}$ of $A_{(\alpha,\beta),0}$ and, by Theorem 1.5, we have

$$\sum_{k=0}^{n} \int_\alpha^\beta a_k(x)[u^{(k)}(x)]^2 \, dx = (\hat{A}_{(\alpha,\beta)}u, u)_{(\alpha,\beta)} = (a[u], u)_{(\alpha,\beta)} = 0 . \tag{4}$$

Thus $\lambda = 0$ is an eigenvalue of $\hat{A}_{(\alpha,\beta)}$ and $u(x)$ is the corresponding eigenfunction. We choose any $\gamma > \beta$. Since

$$0 \geqslant \inf_{\substack{u \in C_0^\infty(\alpha,\beta) \\ \|u\|_{(\alpha,\beta)}=1}} (a[u], u)_{(\alpha,\beta)} > \inf_{\substack{u \in C_0^\infty(\alpha,\gamma) \\ \|u\|_{(\alpha,\gamma)}=1}} (a[u], u)_{(\alpha,\gamma)} \tag{5}$$

there exists a function $v(x) \in C_0^\infty(\alpha, \gamma)$ such that $(a[v], v)_{(\alpha,\gamma)} < 0$. If we continue $v(x)$ as zero elsewhere on $(0, \infty)$ and denote the function so obtained by $v_1(x)$ we have

$$(A_{(1,\infty)}v_1, v_1)_{(1,\infty)} < 0, \quad v_1(x) \in C_0^\infty(1, \infty), \quad \operatorname{supp} v_1 \subseteq (\alpha, \gamma),$$

where $A_{(1,\infty)}$ denotes a self-adjoint extension of $A_{(1,\infty),0}$. If we now choose $X > \gamma$ then, similarly, a function $v_2(x)$ with the properties

$$(A_{(1,\infty)}v_2, v_2)_{(1,\infty)} < 0, \quad v_2(x) \in C_0^\infty(1, \infty), \quad \operatorname{supp} v_2 \subseteq (\gamma, \infty),$$

can be found. If we continue this method and normalize the functions $v_k(x)$, $k = 1, 2. \ldots$, then the result is

$$(A_{(1,\infty)}v_k, v_k)_{(1,\infty)} < 0, \quad v_k(x) \in C_0^\infty(1, \infty), \quad (v_k, v_{k'})_{(1,\infty)} = \begin{cases} 1, & k = k', \\ 0, & k \neq k'. \end{cases}$$

Now suppose that the spectrum of $A_{(1,\infty)}$ has only a finite number of negative points. These points are eigenvalues of $A_{(1,\infty)}$ and the corresponding linearly independent eigenfunctions $u_\nu(x)$, $\nu = 1, \ldots, N$, span an N-dimensional subspace of the Hilbert space. A linear combination $\varphi(x)$ of the functions $v_k(x)$, $k = 1, \ldots$,

$N+1$, with $\| \varphi \|_{(1,\infty)} = 1$, can be constructed which is orthogonal to the N-dimensional subspace. Hence we have

$$(A_{(1,\infty)}\varphi, \varphi)_{(1,\infty)} = \int_{-\infty}^{\infty} \lambda d(E_\lambda \varphi, \varphi) = \int_0^{\infty} \lambda d(E_\lambda \varphi, \varphi) \geqslant 0, \tag{6}$$

where $\{E_\lambda\}_{-\infty<\lambda<\infty}$ denotes the spectral family of $A_{(1,\infty)}$. On the other hand, however, for

$$\varphi(x) = \sum_{k=1}^{N+1} c_k v_k(x)$$

we have

$$(A_{(1,\infty)}\varphi, \varphi)_{(1,\infty)} = \sum_{k=1}^{N+1} | c_k |^2 (A_{(1,\infty)} v_k, v_k) < 0 . \tag{7}$$

Now (6) and (7) contradict each other and hence the spectrum of $A_{(1,\infty)}$ must contain an infinity of negative points.

To prove the converse, we now assume that the negative spectrum of $A_{(1,\infty)}$ is an infinite set. Let $\gamma > 1$ be an arbitrarily chosen point and let $A_{(1,\infty)}$ be decomposed at γ. If $A_{(\gamma,\infty)}$ denotes a self-adjoint extension of $A_{(\gamma,\infty),0}$, then $A_{(1,\infty)}$ and $\hat{A}_{(1,\gamma)} \oplus A_{(\gamma,\infty)}$ are self-adjoint extensions of the operator $A_{\gamma,0}$,

$$A_{\gamma,0} u = a[u] ,$$

$$D(A_{\gamma,0}) = \{u \mid u \in C_0^\infty(1, \infty) \text{ and } u(x) = 0, \ |x - \gamma| < \delta_u, \ \delta_u > 0\} .$$

Since the deficiency indices of $A_{\gamma,0}$ are finite and the spectrum of $A_{(1,\infty)}$ contains an infinity of negative points, the spectrum of $\hat{A}_{(1,\gamma)} \oplus A_{(\gamma,\infty)}$ also possesses an infinity of negative points, by Theorem 1.10. Since, on the negative λ-axis, this spectrum is the union of the corresponding parts of the spectra of $\hat{A}_{(1,\gamma)}$ and $A_{(\gamma,\infty)}$, and since $\sigma(\hat{A}_{(1,\gamma)})$ contains a finite number of negative points only, the set $\sigma(A_{(\gamma,\infty)})$ must possess an infinity of negative points. Therefore there exists a function $u^*(x)$ such that

$$(A_{(\gamma,\infty)} u^*, u^*)_{(\gamma,\infty)} < 0, \quad u^*(x) \in C_0^\infty(\gamma, \Gamma), \quad \Gamma < \infty .$$

From this it follows that the operator $\hat{A}_{(\gamma,\Gamma)}$ possesses at least one negative eigenvalue. If λ_Γ is the least of these eigenvalues we have

$$0>\lambda_\Gamma=\inf_{u\in D(\hat{A}_{(\gamma,\Gamma)})}(\hat{A}_{(\gamma,\Gamma)}u,u)_{(\gamma,\Gamma)}=\inf_{u\in \mathring{W}_2^n(\gamma,\Gamma)}\left(\sum_{k=0}^{n}\int_\gamma^\Gamma a_k(x)|u^{(k)}|^2\,dx\right), \tag{8}$$

where

$$\|u\|_{(\gamma,\Gamma)}=1\,.$$

We show that there is a point Γ_0, $\gamma<\Gamma_0<\Gamma$, such that the least eigenvalue λ_{Γ_0} of $\hat{A}_{(\gamma,\Gamma_0)}$ is equal to zero. By the Variational Principle the value λ_Γ is increasing if Γ moves to the left on the λ -axis with γ fixed. It follows from (8) that λ_Γ is a continuous function of Γ. Finally we show that $\lambda_\Gamma\to\infty$ when $\Gamma\to\gamma$. Since

$$\|u^{(n)}\|^2_{(\gamma,\Gamma)}\geqslant(\Gamma-\gamma)^{-2n}\|u\|^2_{(\gamma,\Gamma)},\quad u\in\mathring{W}_2^n(\gamma,\Gamma)\,,$$

which follows from (3.58) by iteration, and since

$$\|u\|^2_{C^{n-1}(\gamma,\Gamma)}\leqslant C(\Gamma-\gamma)\|u^{(n)}\|^2_{(\gamma,\Gamma)}+C(\Gamma-\gamma)^{1-2n}\|u\|^2_{(\gamma,\Gamma)},$$

which we get from

$$\|u\|^2_{C^{n-1}(0,1)}\leqslant C\|u^{(n)}\|^2_{(0,1)}+C\|u\|^2_{(0,1)}\ \text{(see (A.9))}$$

by the transformation $x=(\Gamma-\gamma)^{-1}(t-\gamma)$, we obtain

$$\sum_{k=0}^{n}\int_\gamma^\Gamma a_k(x)|u^{(k)}|^2\,dx\geqslant\left(\min_{\gamma\leqslant x\leqslant\Gamma}a_n(x)\right)\|u^{(n)}\|^2_{(\gamma,\Gamma)}-C\|u\|^2_{C^{n-1}(\gamma,\Gamma)}$$

$$\geqslant\tfrac{1}{2}\left(\min_{\gamma\leqslant x\leqslant\Gamma}a_n(x)\right)\|u^{(n)}\|^2_{(\gamma,\Gamma)}+\tfrac{1}{2}\left(\min_{\gamma\leqslant x\leqslant\Gamma}a_n(x)\right)(\Gamma-\gamma)^{-2n}\|u\|^2_{(\gamma,\Gamma)}$$

$$-C(\Gamma-\gamma)\|u^{(n)}\|^2_{(\gamma,\Gamma)}-C(\Gamma-\gamma)^{1-2n}\|u\|^2_{(\gamma,\Gamma)},\quad u\in\mathring{W}_2^n(\gamma,\Gamma)\,.$$

From this estimation we have $\lambda_\Gamma\to\infty$ when $\Gamma\to\gamma$. Hence there exists a point $\Gamma_0>\gamma$ such that $\lambda_{\Gamma_0}=0$. The corresponding eigenfunction $u(x)$ of $\hat{A}_{(\gamma,\Gamma_0)}$ is a solution of the differential equation $a[u]=0$ and belongs to the space $\mathring{W}_2^n(\gamma,\Gamma_0)\cap W^{2n}_{2,loc}(\gamma,\Gamma_0)$. Since γ can be chosen arbitrarily we have proved that the differential equation $a[u]=0$ is oscillatory at ∞. This completes the proof of Theorem 5.2.

In this chapter the Sturm Separation Theorem, which we prove now, is also used.

Theorem 5.3
Let the coefficients of the differential equation

$$-(p(x)u')' + q(x)u = \lambda u, \quad x_1 \leqslant x \leqslant x_2 \,, \tag{9}$$

satisfy the conditions

$$p(x) > 0, \quad x \in [x_1, x_2], \quad p(x) \in C^1[x_1, x_2], \quad |q(x)| \leqslant C \,.$$

Then, if x_1 and x_2 are two consecutive zeros of a non-trivial solution $u(x) \in W_2^2[x_1, x_2]$ of (9), each solution $v(x) \in W_2^2[x_1, x_2]$ of (9) has a zero on $[x_1, x_2)$ (or $(x_1, x_2]$).

Proof
From (9) we have

$$0 \equiv (p(x)u')'v - (p(x)v')'u = [p(x)u'v - p(x)uv']'$$

and so

$$p(x_1)u'(x_1)v(x_1) = p(x_2)u'(x_2)v(x_2) \,. \tag{10}$$

From $v(x_1) = 0$ it follows that $v(x_2) = 0$ and vice versa. If, however, the values are both different from zero, then by (10) they must possess different signs because the same is true of $u'(x_1)$ and $u'(x_2)$. Hence there is a zero of $v(x)$ in (x_1, x_2), and Theorem 5.3 is proved.

NON-EXISTENCE OF EIGENVALUES

We now turn to the examination of the spectrum of Sturm-Liouville operators. To prove the non-existence of eigenvalues on intervals of the λ-axis we again make scale transformations on the x-axis. These transformations are now more general than those in the last chapter, and they are used to prove the following lemma.

Lemma 5.4
Let the coefficients of the differential equation

$$-(p(x)u')' + q(x)u = \lambda u, \quad \lambda \text{ real} \tag{11}$$

satisfy the conditions

$$p(x),\, q(x) \in C^1(0, \infty), \quad p(x) > 0,\, q(x) \text{ real-valued} \,.$$

If $\varphi(x)$ and $\rho(x)$ are real-valued and piece-wise smooth, and $\varphi(x)$ has a compact support on $(0, \infty)$, then for a (real-valued) solution $u(x)$ of (11) we have

$$\int_0^\infty [(\rho p' - p\rho')\varphi - \rho p\varphi'](u')^2 \,dx + \int_0^\infty [(\rho(q-\lambda))'\varphi + \rho(q-\lambda)\varphi']u^2 \,dx = 0 . \tag{12}$$

Proof

First we assume that $\rho(x) \in C^1(0, \infty)$ and $\varphi(x) \in C_0^1(0, \infty)$ with

$$\text{supp } \varphi(x) \subseteqq [x_1, x_2], \quad 0 < x_1 < x_2 < \infty .$$

Using the function

$$\vartheta(x) = \begin{cases} 0, x \in (0, \frac{1}{2}x_1 \cup (2x_2, \infty), \\ 1, x \in [x_1, x_2], \\ \in C^\infty(0, \infty), \text{ real-valued} , \end{cases}$$

we define

$$\rho_\vartheta(x) = \begin{cases} \vartheta(x)\rho(x), & 0 < x < \infty, \\ 0, & -\infty < x \leqslant 0 . \end{cases}$$

Then we have

$$\rho_\vartheta(0) = 0 \quad \text{and} \quad |\rho_\vartheta'(x)| \leqslant C, \quad x \in (-\infty, \infty) ,$$

and the function†

$$\xi = \sigma_\epsilon(x) = x + \epsilon\rho_\vartheta(x), \quad 0 < \epsilon < C^{-1} ,$$

transforms the interval $[0, \infty)$ one-to-one onto itself. The inverse transformation is denoted by $x = \sigma_\epsilon^{-1}(\xi)$. If $u(x)$ is a real-valued solution of the differential equation (11), then it follows that from

$$(a[u(x)], \; u[\sigma_\epsilon(x)]\varphi(x)) = \lambda(u(x), u[\sigma_\epsilon(x)]\varphi(x))$$

by means of integration by parts that

† Editor's note: this σ is not to be confused with the notation for a spectrum.

$$\int_0^\infty p(x)u'(x)u'[\rho_\epsilon(x)]\varphi(x)\,dx + \int_0^\infty p(x)u'(x)u[\sigma_\epsilon(x)]\varphi'(x)\,dx$$

$$+\int_0^\infty q(x)u(x)u[\sigma_\epsilon(x)]\varphi(x)\,dx = \lambda\int_0^\infty u(x)u[\sigma_\epsilon(x)]\varphi(x)\,dx\,.^{\ddagger} \tag{13}$$

Similarly from

$$(a[u(x)],\ u[\sigma_\epsilon^{-1}(x)]\varphi[\sigma_\epsilon^{-1}(x)]) = \lambda(u(x),\ u[\sigma_\epsilon^{-1}(x)]\varphi[\sigma_\epsilon^{-1}(x)])$$

we obtain

$$\int_0^\infty p(x)u'(x)u'[\sigma_\epsilon^{-1}(x)]\varphi[\sigma_\epsilon^{-1}(x)]\,dx + \int_0^\infty p(x)u'(x)u[\sigma_\epsilon^{-1}(x)]\varphi'[\sigma_\epsilon^{-1}(x)]\,dx$$

$$+\int_0^\infty q(x)u(x)u[\sigma_\epsilon^{-1}(x)]\varphi[\sigma_\epsilon^{-1}(x)]\,dx = \lambda\int_0^\infty u(x)u[\sigma_\epsilon^{-1}(x)]\varphi[\sigma_\epsilon^{-1}(x)]\,dx\,. \tag{14}$$

In the last equation the transformation $x = \sigma_\epsilon(t)$ is made and then the variable t is replaced by x. This gives

$$\int_0^\infty p[\sigma_\epsilon(x)]\,u'[\sigma_\epsilon(x)]\,u'(x)\varphi(x)[\sigma'_\epsilon(x)]^{-1}\,dx +$$

$$\int_0^\infty p[\sigma_\epsilon(x)]\,u'[\sigma_\epsilon(x)]\,u(x)\varphi(x)[\sigma'_\epsilon(x)]^{-1}\,dx +$$

$$\int_0^\infty q[\sigma_\epsilon(x)]\,u[\sigma_\epsilon(x)]\,u(x)\varphi(x)\sigma'_\epsilon(x)\,dx$$

$$= \lambda\int_0^\infty u[\sigma_\epsilon(x)]\,u(x)\varphi(x)\sigma'_\epsilon(x)\,dx\,. \tag{15}$$

Next, (13) is subtracted from (15), the difference is multiplied by ϵ^{-1} and the limit is calculated when $\epsilon \to 0$. The result is

$$\int_0^\infty [\rho(x)p'(x) - p(x)\rho'(x)](u')^2\varphi(x)\,dx +$$

‡ The differentiations here are with respect to the variable x.

$$\int_0^\infty [\rho(x)p'(x)uu' + \rho(x)p(x)uu'' - \rho(x)p(x)(u')^2]\varphi'(x)\,dx +$$

$$\int_0^\infty [\rho(x)q'(x) + q(x)\rho'(x)]u^2\varphi(x)\,dx = \lambda\int_0^\infty \rho'(x)u^2\varphi(x)\,dx\,. \quad (16)$$

In (16) we use

$$p(x)u''(x) + p'(x)u'(x) = [q(x) - \lambda]u(x)\,,$$

which follows from (11), to obtain

$$\int_0^\infty (\rho p' - p\rho')(u')^2\varphi\,dx + \int_0^\infty [\rho(q-\lambda)u^2 - \rho p(u')^2]\varphi'\,dx +$$

$$\int_0^\infty (\rho q' + q\rho')u^2\varphi\,dx = \lambda\int_0^\infty \rho' u^2\varphi\,dx$$

and hence (12).

If $\varphi(x)$ and $\rho(x)$ are only piece-wise smooth, then we approximate them by C^1-functions in the W^1-norm. Then (12) also follows for piece-wise smooth functions $\varphi(x)$ and $\rho(x)$, and Lemma 5.4 is proved.

Equation (12) is also true for an eigenfunction of a self-adjoint extension of

$$A_0u = a[u],\ \ D(A_0) = C_0^\infty(0,\infty)\,,$$

and the next theorem concerns the existence of such an eigenfunction.

Theorem 5.5

In addition to (2), let the coefficients of the differential expression

$$a[\cdot] = -\frac{d}{dx}p(x)\frac{d}{dx} + q(x),\ \ 0 < x < \infty\,.$$

satisfy

i) $q(x) \in C^1(0,\infty)\,,$

ii) $\limsup_{x\to\infty} q(x) = Q < \infty,\ \ Q \geqslant -\infty\,,$

Let $\lambda > Q$. If there exists a point $x_1 > 0$ and a positive-valued function $\rho(x) \in C^1(0,\infty)$ such that

$$\text{iii)} \quad \limsup_{x \to \infty} x^{-1}(\rho')^2 < \infty ,$$

$$\text{iv)} \quad \limsup_{x \to \infty} | x^{-1} \rho(q - \lambda)| < \infty ,$$

$$\text{v)} \quad \left(\frac{p}{\rho}\right)' \leqslant 0, \quad x_1 \leqslant x < \infty , \tag{17}$$

$$\text{vi)} \quad [\rho(q-\lambda)]' \leqslant 0, \quad x_1 \leqslant x < \infty ,$$

then λ is not an eigenvalue of any self-adjoint extension A of A_0. Also λ belongs to $\sigma_e(A)$.

Proof
For $\varphi(x)$ in (12) we take the function

$$\varphi(x) = \begin{cases} 0, \; x \in (-\infty, x_1] \cup [x_4, \infty) , \\ (x - x_1)/(x_2 - x_1), \quad x \in [x_1, x_2] , \\ 1, \; x \in [x_2, x_3] , \\ (x_4 - x)/(x_4 - x_3), \quad x \in [x_3, x_4] , \\ 0 < x_1 < x_2 < x_3 < x_4 < \infty . \end{cases} \tag{18}$$

The part

$$-\int_{x_1}^{x_2} \rho p (u')^2 \varphi' \, dx + \int_{x_1}^{x_2} \rho(q - \lambda) u^2 \varphi' \, dx$$

of the left-hand side of (12) tends to the value

$$\Omega(x_1) = -\rho(x_1)[u'(x_1)]^2 + \rho(x_1)[q(x_1) - \lambda] u^2(x_1) \tag{19}$$

when $x_2 \to x_1$. Now let x_1 be chosen so large that $\lambda > q(x_1)$, this being possible because $\lambda > Q$. By (12) and (19), we have, as $x_2 \to x_1$,

$$\Omega(x_1) + o(1) + \int_{x_1}^{x_4} \left\{ \rho^2 \left(\frac{p}{\rho}\right)' (u')^2 + [\rho(q-\lambda)]' u^2 \right\} \varphi \, dx - \int_{x_3}^{x_4} [\rho p (u')^2 + \rho(\lambda - q) u^2] \varphi' \, dx = 0 . \tag{20}$$

Let us now suppose that λ is an eigenvalue and $u(x)$ a corresponding eigenfunction. We show that $u(x)$ is oscillatory at ∞. For this purpose we consider the expression

$$\int_0^\infty p(x)[\psi'(x)]^2\,\mathrm{d}x + \int_0^\infty [q(x)-\lambda]\,\psi^2(x)\,\mathrm{d}x$$

where $\psi(x)$ is a suitable function. By (17, v) we can write

$$p(x) = \rho(x)f_1(x),\quad f_1'(x) \leqslant 0,\quad x_1 \leqslant x < \infty, \tag{21}$$

and since, by (17, iv),

$$\rho(x) \leqslant Cx,\quad x_1 \leqslant x < \infty, \tag{22}$$

we obtain

$$p(x) \leqslant Cx,\quad x_1 \leqslant x < \infty. \tag{23}$$

For $\psi(x)$ we choose the function

$$\psi(x) = \begin{cases} 0,\ x \in (-\infty, X_1) \cup (\tfrac{3}{2}X_1, \infty), & X_1 \geqslant x_1\,, \\ 4(x - X_1)/X_1, & x \in [X_1, \tfrac{5}{4}X_1]\,, \\ 4(\tfrac{3}{2}X_1 - x)/X_1, & x \in [\tfrac{5}{4}X_1, \tfrac{3}{2}X_1]\,. \end{cases}$$

Then we have

$$\int_0^\infty p(x)[\psi'(x)]^2\,\mathrm{d}x \leqslant C$$

with a constant C which is independent of X_1. Further we have

$$\int_0^\infty [q(x)-\lambda]\,\psi^2(x)\,\mathrm{d}x \leqslant \tfrac{1}{2}(Q-\lambda)\int_0^\infty \psi^2(x)\,\mathrm{d}x = \tfrac{1}{12}(Q-\lambda)X_1$$

if X_1 is chosen sufficiently large. Altogether, then, we have

$$\int_0^\infty p(x)[\psi'(x)]^2\,\mathrm{d}x + \int_0^\infty [q(x)-\lambda]\,\psi^2(x)\,\mathrm{d}x \leqslant C + \tfrac{1}{12}(Q-\lambda)X_1\,,$$

and it follows that the quadratic form

$$\int_{X_1}^{\frac{3}{2}X_1} (p(x)|\,u'\,|^2 + (q(x) - \lambda)|\,u\,|^2)\,dx\,,$$

with

$$u \in W_2^1(X_1, \tfrac{3}{2}X_1), \quad u(X_1) = u(\tfrac{3}{2}X_1) = 0\,,$$

is negative for $u = \psi$ and a sufficiently large X_1. This implies, however, that the least eigenvalue of $\hat{A}_{(X_1, \frac{3}{2}X_1)}$ is less than λ. If we now move the right-hand end-point left towards the point X_1, then by the Courant Variational Principle we get the situation that the least eigenvalue of the operator $\hat{A}_{(X_1, X_2)}$, $X_1 < X_2 < \frac{3}{2}X_1$, coincides with λ if X_2 is suitably chosen. The corresponding eigenfunction is a solution of the differential equation $a[u] = \lambda u$ with zeros at X_1 and X_2. Now, by Theorem 5.3, it follows that on $[X_1, X_2]$ there is a zero of the eigenfunction $u(x)$ of A which corresponds to λ. Hence $u(x)$ is oscillatory at ∞ because X_1 can be arbitrarily large. Now we postulate that the points x_3 and x_4, which we have used to define the function $\varphi(x)$, coincide with zeros of $u(x)$. So the point x_4 is to be a zero in the interval $[2x_3, 3x_3)$, which we see is possible from our remarks about the zeros of $u(x)$.

We can now estimate the last integral on the left-hand side of (20). First, we have

$$\int_{x_3}^{x_4} \rho(q - \lambda)u^2\varphi'\,dx \to 0 \tag{24}$$

when $x_3 \to \infty$ (where $u(x_3) = 0$) because, by (17, iv), the modulus of the integral is dominated by

$$C\int_{x_3}^{x_4} u^2\,dx\,,$$

which tends to zero since $u \in L_2(0, \infty)$. Next we also show that

$$\int_{x_3}^{x_4} \rho p(u')^2\varphi'\,dx \to 0, \quad x_3 \to \infty\,.$$

By integration by parts we have

$$\lambda\int_{x_3}^{x_4} u^2\rho\varphi'\,\mathrm{d}x = \int_{x_3}^{x_4} a[u]u\rho\varphi'\,\mathrm{d}x$$

$$= \int_{x_3}^{x_4} p(u')^2\rho\varphi'\,\mathrm{d}x + \int_{x_3}^{x_4} puu'\rho'\varphi'\,\mathrm{d}x + \int_{x_3}^{x_4} qu^2\rho\varphi'\,\mathrm{d}x$$

that is,

$$\int_{x_3}^{x_4} \rho(\lambda - q)u^2\varphi'\,\mathrm{d}x = \int_{x_3}^{x_4} \rho p(u')^2\varphi'\,\mathrm{d}x + \int_{x_3}^{x_4} puu'\rho'\varphi'\,\mathrm{d}x\,. \tag{25}$$

The Schwarz inequality gives

$$\left|\int_{x_3}^{x_4} puu'\rho'\varphi'\,\mathrm{d}x\right|^2 \leqslant \left(\int_{x_3}^{x_4} \rho p(u')^2\,|\,\varphi'\,|\,\mathrm{d}x\right)\left(\int_{x_3}^{x_4} p\rho^{-1}u^2(\rho')^2\,|\,\varphi'\,|\,\mathrm{d}x\right). \tag{26}$$

From (17, v) and (17, iii) we obtain

$$p\rho^{-1} \leqslant C,\quad x_3 \leqslant x \leqslant x_4,\quad \text{and}\quad (\rho')^2\,|\,\varphi'\,| \leqslant C,\quad x_3 \leqslant x \leqslant x_4\,,$$

so that in (26) the factor

$$\int_{x_3}^{x_4} p\rho^{-1}u^2(\rho')^2\,|\,\varphi'\,|\,\mathrm{d}x$$

tends to zero when $x_3 \to \infty$. Now by (24), (25), and (26) we have

$$\int_{x_3}^{x_4} \rho p(u')^2\,|\,\varphi'\,|\,\mathrm{d}x = o(1)\left(\int_{x_3}^{x_4} \rho p(u')^2\,|\,\varphi'\,|\,\mathrm{d}x\right)^{\frac{1}{2}} + o(1),\quad x_3 \to \infty\,, \tag{27}$$

and from (27) we obtain

$$\int_{x_3}^{x_4} \rho p(u')^2\varphi'\,\mathrm{d}x = o(1),\quad x \to \infty\,. \tag{28}$$

Therefore the last term on the left-hand side of (20) tends to zero when $x_3 \to \infty$.

By (17 v, vi), the expression

$$N(x_2, x_3) = \int_{x_1}^{x_4}\left\{\rho^2\left(\frac{p}{\rho}\right)'(u')^2 + [\rho(q-\lambda)]'u^2\right\}\varphi\,\mathrm{d}x\,,$$

in which x_4 depends on x_3, is non-positive. Thus (20) gives

$$\Omega(x_1) + N(x_2, x_3) + o(1) = 0, \quad x_2 \to x_1, \quad x_3 \to \infty, \quad N(x_2, x_3) \leqslant 0\,. \tag{29}$$

The term $\Omega(x_1)$ is non-positive, since $\lambda > q(x_1)$, and so, by (29), it can only be zero. This implies, however, that $u(x_1) = u'(x_1) = 0$. Hence $u(x)$, as a solution of the differential equation $a[u] = \lambda u$, must be identically zero. Thus λ cannot be an eigenvalue of A.

Now we show that λ belongs to the essential spectrum of A. Let us assume that $\lambda \notin \sigma_e(A)$. Then λ does not belong to $\sigma_e(\hat{A}_{(x_1,\infty)})$ either, since

$$\sigma_e(\hat{A}_{(x_1,\infty)}) \subseteqq \sigma_e(A)\,.$$

The above proof also shows that λ is not an eigenvalue of $\hat{A}_{(x_1,\infty)}$. Hence λ belongs to the resolvent set of $\hat{A}_{(x_1,\infty)}$ and consequently it is a regular point of $A_{(x_1,\infty),0}$ (see Chapter 1). The deficiency index m of $A_{(x_1,\infty),0}$ is greater than zero, so that by Theorem 1.9 there exists a self-adjoint extension $A_{(x_1,\infty)}$ which has an eigenvalue of multiplicity m at the point λ. However, this is impossible by the first part of the proof. Therefore λ must belong to $\sigma_e(A)$ and Theorem 5.5 is proved.

SPECIAL CASES

We now consider some special cases of the auxiliary function $\rho(x)$.

Theorem 5.6
Let $p(x)$ and $q(x)$ satisfy the following conditions.

i) (2) and $q(x) \in C^1(0, \infty)$,

ii) $\limsup_{x\to\infty} q(x) = Q < \infty, \quad Q \geqslant -\infty$.

1) If there are numbers $x_1 > 0$, $\tau \leqslant 1$, and $\lambda > Q$, and $p(x)$ and $q(x)$ have the form

$$p(x) = x^{\tau} f_1(x), \quad q(x) = \lambda + x^{-\tau} f_2(x),$$

where

$$f_i'(x) \leqslant 0, \quad x_1 \leqslant x < \infty, \quad i = 1, 2, \quad \text{and} \quad \liminf_{x\to\infty} x^{-1} f_2(x) > -\infty,$$

then λ is not an eigenvalue of any self-adjoint extension A of A_0,

$$A_0 u = a[u], \quad D(A_0) = C_0^\infty(0, \infty),$$

and λ belongs to $\sigma_e(A)$.

2) If there is a point x_1 such that $p(x)$ and $q(x)$ are decreasing to the right of x_1, and

$$\liminf_{x \to \infty} x^{-1} q(x) > -\infty, \quad \lim_{x \to \infty} q(x) = Q \geqslant -\infty, \tag{30}$$

then the spectrum of any self-adjoint extension A of A_0 is continuous on (Q, ∞).

3) If for a value τ with $0 < \tau \leqslant 1$ we have

$$xp' - \tau p \leqslant 0, \quad x_1 \leqslant x < \infty, \quad \text{for some} \quad x_1 > 0, \tag{31}$$

$$\limsup_{x \to \infty} (q + \tau^{-1} x q') = \lambda^*, \tag{32}$$

and

$$\liminf_{x \to \infty} x^{\tau - 1} q > -\infty, \tag{33}$$

then the spectrum of A is continuous on the interval

$$(\Lambda, \infty), \quad \Lambda = \max(Q, \lambda^*).$$

4) If, for some τ, $-\infty < \tau < 0$, (31) and (33) hold and

$$\liminf_{x \to \infty} (q + \tau^{-1} x q') = \lambda^*, \quad \lambda^* > Q,$$

then the spectrum of A is continuous on (Q, λ^*).

Proof

1) If we take $\rho(x) = x^\tau$, $\tau \leqslant 1$, then all the conditions in Theorem 5.5 are satisfied, and we have Part (2) immediately

2) This follows from Part (1) with $\tau = 0$ and $\lambda > Q$.

3) If we choose $\lambda > \max(Q, \lambda^*)$, then there is a point $x_1 > 0$ such that both (31) and

$$q(x) + \tau^{-1} x q'(x) < \lambda, \quad x_1 \leqslant x < \infty, \tag{34}$$

hold. From (34) we have

$$\tau x^{\tau-1}(q-\lambda)+x^{\tau}q'=[x^{\tau}(q-\lambda)]'<0,\quad x_1\leqslant x<\infty,$$

and from (31)

$$(x^{-\tau}p)'\leqslant 0,\quad x_1\leqslant x<\infty.$$

Together with (33) all the conditions imposed in (1) are satisfied. Thus λ belongs to $\sigma_e(A)$ but is not an eigenvalue.

4) Let $Q<\lambda<\lambda^*$. Then x_1 can be chosen so large that

$$xp'-\tau p\leqslant 0,$$

and $$q+\tau^{-1}xq'>\lambda,$$

in $[x_1,\infty)$, and hence

$$(x^{-\tau}p)'\leqslant 0,\quad [x^{\tau}(q-\lambda)]'<0,\quad x_1\leqslant x<\infty.$$

Theorem 5.6 is now proved.

We now consider further the special case $\tau=0$ of Theorem 5.6.

Corollary 5.7 [41]
Let the condition (i) of Theorem 5.6 hold and let

ii) $\liminf\limits_{x\to\infty} x^{-1}q(x)>-\infty$.

If there are a point $x_1>0$, a real Λ, and a number $\kappa\leqslant 0$, such that

$$p'\leqslant\kappa p,\quad q'\leqslant\kappa(q-\Lambda),\quad q(x)\leqslant\Lambda$$

in $[x_1,\infty)$, then the spectrum of any self-adjoint extension A of A_0 is continuous on the interval (Λ,∞).

Proof
Suppose that $\lambda>\Lambda$. If there is an eigenfunction $u(x)$ of A corresponding to λ, then $u(x)$ is oscillatory at ∞ as we have shown in the proof of Theorem 5.5. If x_1 is taken as a zero of $u(x)$ and we choose $\rho(x)=1$, then by (29) we have

$$-p(x_1)[u'(x_1)]^2+\int_{x_1}^{x_4}[p'(u')^2+q'u^2]\varphi\,\mathrm{d}x=o(1),\qquad(35)$$

$$x_4=x_4(x_3),\quad x_3\to\infty.$$

Here, φ is as in (18), x_3 and x_4 are zeros of $u(x)$ with $2x_3 \leqslant x_4 < 3x_3$, and we have let $x_2 \to x_1$.

On the other hand, by integration by parts in

$$\lambda(u, u\varphi)_{(x_1,x_4)} = (a[u], u\varphi)_{(x_1,x_4)}$$

we obtain

$$\int_{x_1}^{x_4} [p(u')^2 + (q-\lambda)u^2]\varphi \, dx + \int_{x_3}^{x_4} pu'u\varphi' \, dx = 0 . \tag{36}$$

We show that the second integral on the left-hand side vanishes when $x_3 \to \infty$. We have

$$\lambda(u, u)_{(x_3,x_4)} = (a[u], u)_{(x_3,x_4)} ,$$

and so

$$\int_{x_3}^{x_4} p(u')^2 \, dx = \int_{x_3}^{x_4} (\lambda - q)u^2 \, dx$$

by integration by parts. If this equation is multiplied by $(x_3 - x_4)^{-1}$, we have

$$\int_{x_3}^{x_4} p(u')^2 \varphi' \, dx = \int_{x_3}^{x_4} (\lambda - q)u^2 \varphi' \, dx . \tag{37}$$

From condition (ii) we obtain

$$\left| \int_{x_3}^{x_4} (\lambda - q)u^2 \varphi' \, dx \right| \leqslant C \int_{x_3}^{x_4} u^2 \, dx = o(1), \quad x_3 \to \infty . \tag{38}$$

Hence, by (37) and (38), the first factor on the right-hand side of the inequality

$$\left| \int_{x_3}^{x_4} pu'u\varphi' \, dx \right|^2 \leqslant \left(\int_{x_3}^{x_4} p(u')^2 \,|\, \varphi' \,|\, dx \right) \left(\int_{x_3}^{x_4} pu^2 \,|\, \varphi' \,|\, dx \right) \tag{39}$$

vanishes when $x_3 \to \infty$. The second factor also tends to zero because $p(x)$ is bounded on (x_1, ∞) as a decreasing function. Therefore, by (36) and (39), we obtain

$$\int_{x_1}^{x_4} [p(u')^2 + (q-\lambda)u^2]\varphi \, dx = o(1) .$$

If this equation is multiplied by the real number κ and then subtracted from (35), we obtain

$$-p(x_1)[u'(x_1)]^2 + \int_{x_1}^{x_4} [(p' - \kappa p)(u')^2 + \{q' - \kappa(q - \lambda)\} u^2] \varphi \, dx = o(1),$$

$$x_3 \to \infty .$$

Since, by the hypothesis, the integrand is non-positive, it follows that $u'(x_1) = 0$. Finally, from $u(x_1) = u'(x_1) = 0$ we have $u(x) \equiv 0$. Hence λ is not an eigenvalue of A. The statement that $(\Lambda, \infty) \subseteqq \sigma_e(A)$ can be substantiated as in the proof of Theorem 5.5. Corollary 5.7 is now proved.

If we take $\rho(x) = e^{-\alpha x}$, $\alpha > 0$, in Theorem 5.5 we obtain the following result.

Corollary 5.8
Let the following conditions hold.

i) (2) and $q(x) \in C^1(0, \infty)$,

ii) $\limsup_{x \to \infty} q(x) = Q < \infty$,

iii) $\liminf_{x \to \infty} x^{-1} e^{-\alpha x} q(x) > -\infty, \quad \alpha > 0$,

iv) There is a point $x_1 > 0$, such that

$$\alpha p(x) + p'(x) \leqslant 0, \quad x_1 \leqslant x < \infty,$$

v) $\liminf_{x \to \infty} (q(x) - \alpha^{-1} q'(x)) = \lambda^* > Q$.

Then the spectrum of any self-adjoint extension of A_0,

$$A_0 u = a[u], \quad D(A_0) = C_0^\infty(0, \infty),$$

is continuous on the interval (Q, λ^*).

Proof
If $Q < \lambda < \lambda^*$, then there is a point $x_1 > 0$ such that

$$\lambda < q(x) - \alpha^{-1} q'(x), \quad x_1 \leqslant x < \infty .$$

Therefore with

$$[e^{-\alpha x}(q - \lambda)]' < 0, \quad x_1 \leqslant x < \infty .$$

we have (vi) of Theorem 5.5. The other conditions are also satisfied, and Corollary 5.8 follows.

Examples 5.9

1) Let

$$p(x)=e^{-x},\ q(x)=\sin x,\quad 0<x_1\leqslant x<\infty.$$

With $\Lambda=\sqrt{2}$ we have $q'\leqslant-(q-\Lambda)$, $x\geqslant x_1$, and so by Corollary 5.7 (with $\kappa=-1$) the spectrum of A is continuous on $(\sqrt{2},\infty)$. This fact is no longer true if we take $x_1=0$ and replace $p(x)=e^{-x}$ by $p(x)$ =const >0 since in this case there are gaps in the spectrum.

2) Let

$$p(x)=e^{\kappa x},\quad \kappa<0,\quad q(x)=-cx+b(x),\quad c>0,\quad 0<x_1\leqslant x<\infty,$$

where

$$x^{-1}b(x)\to 0,\ xb'(x)\to 0,\quad x\to\infty.$$

For each Λ, the inequalities

$$p'(x)\leqslant\kappa p(x),\ q'\leqslant\kappa(q-\Lambda),\quad q(x)\leqslant\Lambda,$$

hold for sufficiently large x and so by Corollary 5.7 the spectrum of A is continuous on $(-\infty,\infty)$.

3) Let

$$p(x)=x^{\alpha_1}e^{-\beta x},\quad q(x)=-x^{\alpha_2}e^{\beta x},\quad 0<x_1\leqslant x<\infty,$$

$$0<\alpha_1+\alpha_2\leqslant 1,\quad 0<\beta<\infty.$$

Using Theorem 5.5, we now have $Q=-\infty$ and $\rho(x)=p(x)$. The spectrum of A is continuous on $(-\infty,\infty)$.

We now consider the special case

$$\liminf_{x\to\infty}\ q(x)>-\infty.$$

The conditions on the coefficients in Theorem 5.5 can then be weakened. As a preliminary we prove

Lemma 5.10
The operator

$$A_0u=-(p(x)u')'+q(x)u,\quad D(A_0)=C_0^\infty(-\infty,\infty),$$

is essentially self-adjoint if the following conditions are satisfied.

$$p(x) \in W^1_{2,\mathrm{loc}}(-\infty, \infty), \quad q(x) \in L_{2,\mathrm{loc}}(-\infty, \infty),$$

$$p(x) > 0, \quad -\infty < x < \infty, \quad \inf_{-\infty < x < \infty} q(x) > -\infty .$$

Proof

Let a number c be chosen so large that

$$q(x) + c \geqslant 1, \quad -\infty < x < \infty .$$

To establish the essential self-adjointness of A_0 we show that the range of $A_0 + cE$ is dense in H. If $v \in H$ is such that

$$((A_0 + cE)u, v) = 0, \quad u \in D(A_0), \tag{40}$$

it follows that v belongs to $D(A_0^*)$. From Theorem 2.1 we have

$$D(A_0^*) \subseteqq W^2_{2,\mathrm{loc}}(-\infty, \infty)$$

and so (40) gives

$$(u, a[v] + cv) = 0, \quad u \in D(A_0) .$$

Hence we obtain

$$-(p(x)v')' + (q(x) + c)v = 0 \tag{41}$$

and the lemma is proved if we show that $v(x) \equiv 0$ is the only solution of the differential equation (41) in $L_2(-\infty, \infty)$. To do this we argue as follows assuming, as we may, that $v(x)$ is real-valued. We have

$$0 = \int_s^t [-(p(x)v')' + (q(x) + c)v]\, v \, dx$$

$$= \int_s^t [p(x)(v')^2 + (q(x) + c)v^2] \, dx$$

$$+ \tfrac{1}{2}[p(s)(v^2)'(s) - p(t)(v^2)'(t)] \tag{42}$$

$$\geqslant \int_s^t v^2 \, dx + \tfrac{1}{2}[p(s)(v^2)'(s) - p(t)(v^2)'(t)] .$$

If $v(x)$ is a non-trivial $L_2(-\infty, \infty)$ solution of (41), then there exist numbers s and t such that

$$p(s)(v^2)'(s) > 0 \quad \text{and} \quad p(t)(v^2)'(t) < 0 .$$

Then from (42) we obtain

$$\int_s^t v^2(x)\, \mathrm{d}x < 0 .$$

This, however, is impossible, and we must have $v(x) \equiv 0$. If $\bar{A}_0$ denotes the closure of A_0, then with

$$R(\bar{A}_0 + cE) = \overline{R(A_0 + cE)} = H = L_2(-\infty, \infty)$$

the self-adjointness of $\bar{A}_0$ is proved.

We can now prove

Theorem 5.11

Let the following conditions hold.

i) (2) and $q(x) \in C^1(0, \infty)$,

ii) $\limsup\limits_{x\to\infty} q(x) = Q < \infty, \quad \liminf\limits_{x\to\infty} q(x) > -\infty$.

If there are a point $x_1 > 0$, a number $\lambda > Q$, and a positive function $\rho(x) \in C^1(0, \infty)$ such that

iii) $\limsup\limits_{x\to\infty} x^{-1} \rho(x) < \infty$,

iv) $\left(\dfrac{p}{\rho}\right)' \leqslant 0, \quad x_1 \leqslant x < \infty$,

v) $[\rho(q - \lambda)]' \leqslant 0, x_1 \leqslant x < \infty$,

then λ is not an eigenvalue of any self-adjoint extension A of A_0,

$$A_0 u = a[u], \quad D(A_0) = C_0^\infty(0, \infty),$$

and λ belongs to $\sigma_e(A)$.

Proof

We have only to show that under the new conditions the term

$$\int_{x_3}^{x_4} [\rho p(u')^2 + \rho(\lambda - q)u^2]\varphi' \, dx \tag{43}$$

in (20) tends to zero when $x_3 \to \infty$. We now take $x_4 = 2x_3$. Then we have

$$\left| \int_{x_3}^{2x_3} \rho\lambda u^2 \varphi' \, dx \right| \leqslant C \int_{x_3}^{2x_3} u^2 \, dx \to 0, \quad x_3 \to \infty. \tag{44}$$

By (ii), without loss of generality we may assume that $q(x)$ is positive on $[x_1, \infty)$. Then, by (iii), we obtain

$$\left| \int_{x_3}^{2x_3} [\rho p(u')^2 - \rho q u^2]\varphi' \, dx \right| \leqslant C \int_{x_3}^{2x_3} [p(u')^2 + qu^2] \, dx \, . \tag{45}$$

To show that the right-hand side of (45) tends to zero as $x_3 \to \infty$ we prove now that the integral

$$\int_{x_1}^{\infty} [p(u')^2 + qu^2] \, dx$$

converges, and we use Lemma 5.10. First, the functions $p(x)$, $q(x)$, and $u(x)$ are continued from x_1 to the left as functions $\tilde{p}(x)$, $\tilde{q}(x)$, and $\tilde{u}(x)$ in such a manner that

$$\tilde{p}(x) \begin{cases} = p(x), \; x_1 \leqslant x < \infty, \\ > 0, \; -\infty < x < \infty, \\ \in C^1(-\infty, \infty), \end{cases} \qquad \tilde{q}(x) \begin{cases} = q(x), \; x_1 \leqslant x < \infty, \\ > -C, \; -\infty < x < \infty, \\ \in C^1(-\infty, \infty), \end{cases}$$

and

$$\tilde{u}(x) \begin{cases} = u(x), \; x_1 \leqslant x < \infty, \\ = 0, \; -\infty < x \leqslant x_1 - 1, \\ \in C^2(-\infty, \infty). \end{cases}$$

Then $\tilde{u}(x)$ belongs to the domain $D(\tilde{A}_0^*)$ of the adjoint operator of

$$\tilde{A}_0 w = -(\tilde{p}(x)w')' + \tilde{q}(x)w, \quad D(\tilde{A}_0) = C_0^\infty(-\infty, \infty).$$

By Lemma 5.10 the operator $\tilde{A}_0$ is essentially self-adjoint, and so we have

$$\tilde{A}_0^* = (\bar{\tilde{A}}_0)^* = \bar{\tilde{A}}_0$$

where $\bar{\tilde{A}}_0$ denotes the closure of $\tilde{A}_0$. Hence $\tilde{u} \in D(\bar{\tilde{A}}_0)$. Since $\bar{\tilde{A}}_0$ is identical to the self-adjoint operator defined by the form

$$\int_{-\infty}^{\infty} (\tilde{p}(x)w'\bar{z}' + \tilde{q}(x)w\bar{z})\,dx, \quad w, z \in C_0^\infty(-\infty, \infty),$$

the integral

$$\int_{-\infty}^{\infty} [\tilde{p}(x)(\tilde{u}')^2 + \tilde{q}(x)\tilde{u}^2]\,dx$$

must converge. Hence

$$\int_{x_3}^{2x_3} [\tilde{p}(\tilde{u}')^2 + \tilde{q}\tilde{u}^2]\,dx = \int_{x_3}^{2x_3} [p(u')^2 + qu^2]\,dx$$

tends to zero when $x_3 \to \infty$.

From (44) and (45) it follows now that the integral (43) also vanishes when $x_3 \to \infty$. Thus we have (29) again, and Theorem 5.11 follows

By specializing Theorem 5.11, we can state a corollary.

Corollary 5.12

Let (i) and (ii) of Theorem 5.11 hold. Also, for a number $\lambda > Q$, let either

1) $\limsup_{x\to\infty} x^{-1}p(x) < \infty, \quad [p(q-\lambda)]' \leqslant 0, \quad x_1 \leqslant x < \infty,$

or

2) $[p(q-\lambda)]' \geqslant 0, \quad x_1 \leqslant x < \infty.$

Then λ is not an eigenvalue of any self-adjoint extension A of A_0 and λ belongs to $\sigma_e(A)$.

Proof

Under the first set of conditions (1), the conditions in Theorem 5.11 are satisfied if we take $\rho(x) = p(x)$. Under the second set of conditions, we define instead

$$\rho(x) = \begin{cases} [\lambda - q(x_1)]^{-1}, 0 < x < x_1, \\ \\ [\lambda - q(x)]^{-1}, x_1 \leqslant x < \infty, \end{cases}$$

where x_1 is chosen so large that

$$\lambda - q(x) \geqslant \delta > 0, \quad x_1 \leqslant x < \infty.$$

Corollary 5.13
Let (i) and (ii) of Theorem 5.11 hold. Then

1) If there exists a point $x_1 > 0$ such that

$$p'(x) \geqslant 0, \quad q'(x) \leqslant 0, \quad p(x) \leqslant Cx, \quad x_1 \leqslant x < \infty,$$

then the spectrum of A is continuous on the interval

$$(q_\infty, \infty), \quad q_\infty = \lim_{x \to \infty} q(x)$$

2) If there exists a point $x_1 > 0$ such that

$$p'(x) \leqslant 0, \quad q'(x) \geqslant 0, \quad x_1 \leqslant x < \infty,$$

then again the spectrum of A is continuous on (q_∞, ∞).

Proof
In the first case we have

$$[p(q - \lambda)]' \leqslant 0, \quad X_\lambda \leqslant x < \infty,$$

for each $\lambda > q_\infty$, if $X_\lambda (\geqslant x_1)$ is chosen suffuciently large. In the second case we have

$$[p(q - \lambda)]' \geqslant 0, \quad x_1 \leqslant x < \infty,$$

for each $\lambda > q_\infty$. Hence the corollary follows directly from Corollary 5.12.

Note 5.14
If we take $\rho(x) = 1$ in Theorem 5.12, then we obtain the result that the spectrum of A is continuous on the interval (q_∞, ∞) if there exists a point $x_1 > 0$ such that

$$p'(x) \leqslant 0, \quad q'(x) \leqslant 0, \quad x_1 \leqslant x < \infty, \quad \lim_{x \to \infty} q(x) = q_\infty > -\infty.$$

This is basically a theorem of Kreith [28], see also [59].

We go on to discuss the case

$$\limsup_{x\to\infty} q(x) = Q < \infty, \quad \liminf_{x\to\infty} q(x) > -\infty, \quad \rho(x) = x\,.$$

If we put $x_4 = 2x_3$ and let $x_2 \to x_1$ in (20), we obtain

$$\Omega(x_1) + \int_{x_1}^{2x_3} [(xp' - p)(u')^2 + (q - \lambda + xq')u^2]\varphi\,\mathrm{d}x = o(1), \quad x_3 \to \infty, \tag{46}$$

where now

$$\varphi(x) = \begin{cases} 1, & x_1 \leqslant x \leqslant x_3\,, \\ (2x_3 - x)/x_3, & x_3 \leqslant x \leqslant 2x_3\,. \end{cases}$$

We again consider the case where the eigenfunction $u(x)$ is oscillatory at ∞ and now postulate that x_1 is a zero of $u(x)$. Then from (19) we have

$$\Omega(x_1) = -x_1 p(x_1)[u'(x_1)]^2\,.$$

By integration by parts in

$$(a[u], u\varphi)_{(x_1, 2x_3)} = \lambda(u, u\varphi)_{(x_1, 2x_3)}$$

we obtain

$$\int_{x_1}^{2x_3} p(u')^2\varphi\,\mathrm{d}x + \int_{x_1}^{2x_3} (q - \lambda)u^2\varphi\,\mathrm{d}x + \int_{x_1}^{2x_3} pu'u\varphi'\,\mathrm{d}x = 0\,.$$

This equation is multiplied by $1 - \kappa$, where κ is a real parameter, and is then added to (46). The result is

$$-x_1 p(x_1)[u'(x_1)]^2 + \int_{x_1}^{2x_3} (xp' - \kappa p)(u')^2\varphi\,\mathrm{d}x$$

$$+ \int_{x_1}^{2x_3} [(2 - \kappa)(q - \lambda) + xq']\,u^2\varphi\,\mathrm{d}x + (1 - \kappa)\int_{x_3}^{2x_3} puu'\varphi'\,\mathrm{d}x \to 0, \quad x_3 \to \infty\,. \tag{47}$$

We use (47) to prove the next theorem.

Theorem 5.15 [40]
Let the following conditions hold.

i) (2) and $q(x) \in C^1(0, \infty)$,

ii) $\liminf_{x \to \infty} q(x) > -\infty$.

Also, let a point $x_1 > 0$ and a number $\kappa \leqslant 2$ exist such that

$$(x^{-\kappa} p(x))' \leqslant 0 \quad \text{and} \quad (x^{2-\kappa} q(x))' \leqslant 0, \quad x_1 \leqslant x < \infty, \tag{48}$$

In addition, let $q(x) \to 0$ as $x \to \infty$ when $\kappa = 2$. Then no self-adjoint extension A of

$$A_0 u = a[u], \quad D(A_0) = C_0^\infty(0, \infty),$$

possesses an eigenvalue λ with

$$\lambda > \Lambda = \tfrac{1}{4} \lim_{x \to \infty} x^{-2} p(x),$$

and we have

$$[\Lambda, \infty) \subseteqq \sigma_e(A).$$

Proof
From (48) we have

$$p(x) \leqslant Cx^2, \quad x_1 \leqslant x < \infty.$$

Then, for an eigenfunction $u(x)$ of A which is oscillatory at ∞, we have

$$\left| \int_{x_3}^{2x_3} p u u' \varphi' \, \mathrm{d}x \right|^2 \leqslant \left(\int_{x_3}^{2x_3} p (u')^2 \, \mathrm{d}x \right) \left(\int_{x_3}^{2x_3} p u^2 (\varphi')^2 \, \mathrm{d}x \right)$$
$$\leqslant C \left(\int_{x_3}^{2x_3} p (u')^2 \, \mathrm{d}x \right) \left(\int_{x_3}^{2x_3} u^2 \, \mathrm{d}x \right).$$

The two integrals on the right-hand side both tend to zero when $x_3 \to \infty$, as in the proof of Theorem 5.11. Hence

$$\int_{x_3}^{2x_3} p u u' \varphi' \, \mathrm{d}x = o(1), \quad x_3 \to \infty,$$

and from (47) we have

$$-x_1 p(x_1)[u'(x_1)]^2 + \int_{x_1}^{2x_3} (xp' - \kappa p)(u')^2 \varphi \, dx +$$

$$\int_{x_1}^{2x_3} [(2 - \kappa(q - \lambda) + xq']u^2 \varphi \, dx = o(1), \quad x_3 \to \infty. \tag{49}$$

Also, from (48) we can write the coefficients in the form

$$p(x) = x^{\kappa} f_1(x), \quad q(x) = x^{\kappa-2} f_2(x), \quad f_i'(x) \leqslant 0, \quad x_1 \leqslant x < \infty, \quad i = 1, 2. \tag{50}$$

First, we consider the case $\kappa = 2$ and

$$\lim_{x \to \infty} f_1(x) = f_1(\infty) > 0.$$

By Theorem 2.22, the spectrum of $A_{(x_1, \infty)}$ is the interval

$$[\Lambda, \infty), \quad \Lambda = \tfrac{1}{4} f_1(\infty) = \tfrac{1}{4} \lim_{x \to \infty} x^{-2} p(x),$$

and hence $[\Lambda, \infty) \subseteqq \sigma_e(A)$. If now $\lambda > \Lambda$, by Theorem 5.2, the differential equation $a[u] = \lambda u$ is oscillatory at ∞ and, by Theorem 5.3, any eigenfunction $u(x)$ belonging to λ is oscillatory at ∞. Hence (49) holds in the form

$$-x_1 p(x_1)[u'(x_1)]^2 + \int_{x_1}^{2x_3} (xp' - 2p)(u')^2 \varphi \, dx + \int_{x_1}^{2x_3} xq'u^2 \varphi \, dx = o(1), \quad x_3 \to \infty. \tag{51}$$

Since the integrands are non-positive, by making $x_3 \to \infty$ we get $u'(x_1) = 0$. Then from $u(x_1) = u'(x_1) = 0$ we obtain $u(x) \equiv 0$. Thus any $\lambda > \Lambda$ is not an eigenvalue and the theorem is proved in the case $\kappa = 2, f_1(\infty) > 0$.

Now let $\kappa = 2$ and $f_1(\infty) = 0$. Then $\Lambda = 0$ and by supposition we have $\lim_{x \to \infty} q(x) = 0$. If a number $\eta > 0$ is chosen arbitrarily, then there is a point $x_\eta > x_1$ such that

$$p(x) \leqslant \eta x^2, \quad q(x) \leqslant \eta, \quad x_\eta \leqslant x < \infty. \tag{52}$$

By Note 2.21, the spectrum of the operator $\hat{B}_{(x_\eta, \infty)}$,

$$B_{(x_\eta, \infty), 0} u = -\eta(x^2 u')' + \eta, \quad D(B_{(x_\eta, \infty), 0}) = C_0^\infty(x_\eta, \infty),$$

is the interval $[\frac{5}{4}\eta, \infty)$, so that by the Courant Variational Principle (Theorem 1.6), there must be an infinite number of points of the spectrum of $\hat{A}_{(x_1,\infty)}$ to the left of 2η. Since $\eta > 0$ is arbitrary, the differential equation $a[u] = \lambda u$ is oscillatory at ∞ for every $\lambda > 0$. Therefore the eigenfunction $u(x)$ belonging to λ is oscillatory at ∞. Now from (51) we can again infer that $u(x)$ must vanish identically. Hence there is no positive eigenvalue. That $[0, \infty) \subseteqq \sigma_e(A)$ can be shown as in the proof of Theorem 5.5.

In the case where $\kappa < 2$ we likewise have $\Lambda = 0$ and

$$\limsup_{x\to\infty} q(x) \leqslant 0 .$$

We can now argue as in the case $\kappa = 2$ and $f_1(\infty) = 0$. Theorem 5.15 is now proved.

Examples 5.16

1) If (i) of Theorem 5.15 and

$$p(x) = 1, \quad q'(x) \leqslant 0, \quad x_1 \leqslant x < \infty, \quad \lim_{x\to\infty} q(x) = 0 , \tag{53}$$

all hold, then there is no positive eigenvalue of any self-adjoint extension A of A_0. Since in this case we have $\sigma_e(A) = [0, \infty)$ there is no eigenvalue in the interior of the essential spectrum. This example applies where $\kappa = 2$ in Theorem 5.15. Next we show that, when $\kappa < 2$ and $q(x)$ may be negative for large x, it is possible for eigenvalues to occur in the interior of the essential spectrum. In [7] the asymptotic behaviour of solutions of the differential equation

$$u'' + [1 + h(x) \sin(2x)]u = 0, \quad x > 0 ,$$

is described. Assuming that

$$h(x) \in C^1(0, \infty) \text{ and real-valued}, \quad h'(x) < 0, \quad x \geqslant x_0 > 0, \quad \int_{x_0}^{\infty} h^2(x)\,dx < \infty ,$$

there are solutions $u_i(x)$, $i = 1, 2$, which behave asymptotically as

$$u_1(x) = \rho(x)[\cos x + o(1)] , \; u_2(x) = [\rho(x)]^{-1} [\sin x + o(1)], \quad x \to \infty,$$

$$\rho(x) = \exp\left(\tfrac{1}{4}\int_{x_0}^{x} h(t)\,dt\right) .$$

Therefore the differential equation

$$-u'' - 4x^{-1} \sin(2x)u = u, \quad 0 < x < \infty ,$$

has solutions which behave asymptotically as

$$u_1(x) = x[\cos x + o(1)], \quad u_2(x) = x^{-1}[\sin x + o(1), \quad x \to \infty .$$

Here, $u_2(x)$ belongs to $L_2(0, \infty)$ and there is a definite α, $0 \leqslant \alpha < \pi$, such that

$$u_2(0)\cos\alpha - u_2'(0)\sin\alpha = 0 . \tag{54}$$

Hence $u_2(x)$ is an eigenfunction of the self-adjoint extension $A^{(\alpha)}$ of

$$A_0 u = -u'' - 4x^{-1}\sin(2x)u, \quad D(A_0) = C_0^\infty(0, \infty) ,$$

which corresponds to the boundary condition (54). The eigenvalue is $\lambda = 1$, which lies in the interior of the essential spectrum $\sigma_e(A^{(\alpha)}) = [0, \infty)$. The application of Theorem 5.15 to this example gives the result that there is no eigenvalue of any self-adjoint extension of A_0 in the interval $(4, \infty)$, as can be seen from the following. The functions

$$p(x) = 1, \quad q(x) = -4x^{-1}\sin(2x) - c$$

satisfy the conditions of Theorem 5.15 with $\kappa = 0$ for each $c > 4$. Hence, for these values of c, the self-adjoint extensions of

$$(A_0 - cE)u = -u'' + (4x^{-1}\sin(2x) + c)u, \quad u \in C_0^\infty(0, \infty) ,$$

have no positive eigenvalue. Thus no self-adjoint extension of A_0 has an eigenvalue greater than 4.

Other examples of functions $q(x)$, for which there are eigenvalues in the interior of the essential spectrum, can be found in [12] or [13] for example.

2) We consider the special case

$$p(x) = xf_1(x), \quad q(x) = x^{-1}f_2(x), \quad f_i(x) \in C^1(0, \infty), \quad i = 1, 2,$$
$$f_1(x) > 0, \quad 0 < x < \infty; \quad f_1'(x) \leqslant 0, \quad f_2'(x) \leqslant \gamma, \quad q(x) \geqslant -C,$$
$$x \geqslant x_1 > 0 .$$

Then the spectrum of the relevant self-adjoint operators is continuous on (γ, ∞). This follows from Theorem 5.15 with $\kappa = 1$.

3) Let

$$p(x) = x^\alpha, \quad \alpha < 2, \quad q(x) = \sin[\log x] , \quad x \geqslant x_1 > 0 .$$

Then the spectrum is continuous on (γ, ∞), where

$$\gamma = \max_{0 \leqslant \varphi \leqslant 2\pi} [\sin \varphi + (2 - \alpha)^{-1} \cos \varphi] ,$$

which follows from Theorem 5.15 with $\kappa = \alpha$.

Note 5.17

In the special case $p(x) = 1$ there are in the literature other results on the non-existence of eigenvalues on intervals, which are more general than the comparable ones of this chapter. As far as we know† the most general results are stated by Eastham [11] as follows. Let

$$q(x) = r(x) + s(x) \text{ where } s(x) \in C^1 [0, \infty) \text{ and } \lim_{x \to \infty} r(x) = \lim_{x \to \infty} s(x) = 0 .$$

If the numbers

$$K = \limsup_{x \to \infty} x \,|\, r(x) \,|, \quad 0 \leqslant K < \infty ,$$

$$L = \limsup_{x \to \infty} x s'(x), \quad 0 \leqslant L < \infty ,$$

$$\Lambda = \tfrac{1}{2} [K^2 + L + K(K^2 + 2L)^{1/2}] ,$$

are defined, then the spectrum of every self-adjoint extension of

$$A_0 u = -u'' + q(x) u, \quad D(A_0) = C_0^\infty (0, \infty) ,$$

is continuous on (Λ, ∞). In the special case $s(x) = 0$ this result was stated by Wallach[57].

In another theorem [12] the functions $r(x)$ and $s(x)$ are subject to integral conditions. Under the conditions

$$s(x) \in C^1 [0, \infty), \,|\, s(x) \,|\, \leqslant C, 0 \leqslant x < \infty, \quad \limsup_{x \to \infty} s(x) = 0 ,$$

there is no eigenvalue $\lambda > \Lambda_1$ where

$$\Lambda_1 = \tfrac{1}{2} [K^2 + 2L + K(K^2 + 4L)^{1/2}] ,$$

†Editor's Note: see also the paper of Atkinson and Everitt, *Proc. Roy. Soc. Edinburgh* (A) 80 (1978) 57–66.

$$K = \limsup_{x\to\infty} (\log x)^{-1} \int_0^x |r(t)|\, \mathrm{d}t, \quad 0 \leqslant K < \infty,$$

$$L = \limsup_{x\to\infty} (\log x)^{-1} \int_0^x \max[s'(t), 0]\, \mathrm{d}t, \quad 0 \leqslant L < \infty.$$

In the case $s(x) = 0$ this result goes back to Borg [5], and the case $r(x) = 0$ was considered by Weidmann [59].

Here, we also refer to a paper of Hinton [19] where the continuity of the spectrum on $(0, \infty)$ is proved for (1) under certain integral conditions on $p(x)$ and $q(x)$ and their first two derivatives.

In the previous theorems the behaviour of the coefficients $p(x)$ and $q(x)$ was prescribed for large x in such a way that statements could be made about the non-existence of eigenvalues. The behaviour of the coefficients in the neighbourhood of $x = 0$, on the other hand, was not restricted, apart from the general conditions (2) and $q(x) \in C^1(0, \infty)$. In the following theorem we interchange the rôles of the boundary points ∞ and 0 and formulate sufficient conditions on $p(x)$ and $q(x)$ in the neighbourhood of $x = 0$ to obtain a similar statement about the non-existence of eigenvalues.

Theorem 5.18

In addition to (2), let

i) $q(x) \in C^1(0, \infty)$,

ii) $\liminf_{x\to 0} q(x) > -\infty$.

If there are a point $x_0 > 0$ and a number $\kappa \geqslant 2$ such that

$$(x^{-\kappa} p(x))' \geqslant 0 \text{ and } (x^{2-\kappa} q(x))' \geqslant 0, \quad 0 < x \leqslant x_0, \tag{55}$$

with $\lim_{x\to 0} q(x) = 0$ in the case $\kappa = 2$, then no self-adjoint extension A of A_0 has an eigenvalue λ with

$$\lambda > \Lambda = \tfrac{1}{4} \lim_{x\to 0} x^{-2} p(x),$$

and $[\Lambda, \infty) \subseteqq \sigma_e(A)$.

Proof

The proof is similar to that of Theorem 5.15 and it is therefore omitted.

We return to the case $\tau = 0$ of Theorem 5.6. For the special case $p(x) = 1$ there is a corresponding result by Hartman and Wintner [16] where (30) is replaced by the weaker condition (56) (below). In the next theorem, we consider this situation, also with a general $p(x)$.

Theorem 5.19
Let the following conditions hold.

i) (2) and $q(x) \in C^1(0, \infty)$,

$$\text{ii)} \int_1^\infty [q(x)]^{-1}\, dx = -\infty. \tag{56}$$

If the coefficients $p(x)$ and $q(x)$ are both decreasing to the right of a point $x_1 > 0$ and

$$q(x) \to -\infty, \quad x \to \infty,$$

then the spectrum of any self-adjoint extension of A_0 is continuous on the whole axis $(-\infty, \infty)$.

Proof
If in (12) we take $\rho(x) = 1$, then we obtain

$$\int_0^\infty (p'\varphi - p\varphi')(u')^2\, dx + \int_0^\infty [q'\varphi + (q-\lambda)\varphi']u^2\, dx = 0 \tag{57}$$

for any (real-valued) eigenfunction $u(x)$ belonging to λ. We take $\varphi(x)$ to be as in (18) on the interval $[0, x_3]$, but on $[x_3, \infty)$ we define $\varphi(x)$ anew. To do this we show that the distance between consecutive zeros of the eigenfunction $u(x)$, which oscillates at ∞, forms a decreasing sequence. Let x_1 be chosen so large that

$$\lambda - q(x) > 0, \quad q(x) < 0, \quad x_1 \leqslant x < \infty,$$

and, as before, let $u(x_1) = u(x_3) = 0$. Let x_μ, $\mu = 4, 5, \ldots$, be the succeeding zeros of $u(x)$. Since $p(x)$ and $q(x)$ are decreasing on $[x_3, \infty)$ we have

$$\inf_{v \in C_0^\infty(I)} \int_I [p\,|v'|^2 + q\,|v|^2]\, dx \geqslant \inf_{v \in C_0^\infty(I')} \int_{I'} [p\,|v'|^2 + q\,|v|^2]\, dx,$$

where $\|v\| = 1$, $I = (x_3, x_4)$, and $I' = (x_4, 2x_4 - x_3)$. From this, we obtain $x_5 \leqslant 2x_4 - x_3$, and hence

$$x_5 - x_4 \leqslant x_4 - x_3.$$

Generally, then, we have

$$x_{\mu+1} - x_\mu \leqslant x_\mu - x_{\mu-1}, \quad \mu = 4, 5, \ldots. \tag{58}$$

Now we define

$$\tilde{\varphi}'(x) = \max_{x_\mu \leqslant x \leqslant x_{\mu+1}} [q(x)]^{-1} = [q(x_{\mu+1})]^{-1}, \ x_\mu \leqslant x \leqslant x_{\mu+1}, \ \mu = 3, 4, \ldots \tag{59}$$

and

$$\tilde{\varphi}(x) = 1 + \int_{x_3}^{x} \tilde{\varphi}'(t)\,\mathrm{d}t, \ x_3 \leqslant x < \infty,$$

Then

$$\begin{aligned}\tilde{\varphi}(x_M) &= 1 + \sum_{\mu=3}^{M-1} [q(x_{\mu+1})]^{-1}(x_{\mu+1} - x_\mu) \\ &\leqslant 1 + \sum_{\mu=4}^{M-1} [q(x_\mu)]^{-1}(x_{\mu+1} - x_\mu) \leqslant 1 + \int_{x_4}^{x_{M+1}} [q(x)]^{-1}\,\mathrm{d}x\,.\end{aligned} \tag{60}$$

by (58). Since the right-hand side tends to $-\infty$ when $M \to \infty$ there exists an index $M \geqslant 3$ such that

$$\tilde{\varphi}(x_M) > 0, \ \tilde{\varphi}(x_{M+1}) \leqslant 0\,. \tag{61}$$

Altogether, our new definition of $\varphi(x)$ is

$$\varphi(x) = \begin{cases} 0, & 0 < x \leqslant x_1, \\ (x - x_1)/(x_2 - x_1), & x_1 \leqslant x \leqslant x_2, \\ 1, & x_2 \leqslant x \leqslant x_3, \\ \tilde{\varphi}(x), & x_3 \leqslant x \leqslant x_M, \\ \tilde{\varphi}(x_M)(x_{M+1} - x)/(x_{M+1} - x_M), & x_M \leqslant x \leqslant x_{M+1}, \\ 0, & x_{M+1} \leqslant x < \infty. \end{cases}$$

If this function is substituted in (57) and the limit $x_2 \to x_1$ taken, we obtain

$$\begin{aligned}&-p(x_1)[u'(x_1)]^2 + \int_{x_1}^{x_{M+1}} [p'(u')^2 + q'u^2]\varphi\,\mathrm{d}x \\ &+ \int_{x_3}^{x_{M+1}} [(q - \lambda)u^2 - p(u')^2]\varphi'\,\mathrm{d}x = 0\,.\end{aligned} \tag{62}$$

We estimate the second integral in (62). As $x_3 \to \infty$, we have

$$\left| \int_{x_3}^{x_{M+1}} q u^2 \varphi' \, dx \right| \leqslant \sum_{\mu=3}^{M} \int_{x_\mu}^{x_{\mu+1}} | q | u^2 | q(x_{\mu+1}) |^{-1} \, dx$$

$$\leqslant \int_{x_3}^{x_{M+1}} u^2 \, dx = o(1) ,$$

and so

$$\int_{x_3}^{x_{M+1}} (q - \lambda) u^2 \varphi' \, dx = o(1) . \tag{63}$$

Also, integrating by parts in

$$\lambda(u, u)_{(x_\mu, x_{\mu+1})} = (a[u], u)_{(x_\mu, x_{\mu+1})}$$

we obtain

$$\int_{x_\mu}^{x_{\mu+1}} p(u')^2 \, dx = \int_{x_\mu}^{x_{\mu+1}} (\lambda - q) u^2 \, dx, \quad \mu = 3, \ldots, M ,$$

and hence

$$\int_{x_3}^{x_{M+1}} p(u')^2 \varphi' \, dx = \int_{x_3}^{x_{M+1}} (\lambda - q) u^2 \varphi' \, dx . \tag{64}$$

Summarizing (62), (63), and (64) we have

$$-p(x_1)[u'(x_1)]^2 + N(x_3) = o(1), \quad x_3 \to \infty , \tag{65}$$

where $N(x_3)$ is non-positive. From this we again obtain $u'(x_1) = 0$. Finally, since $u(x_1) = u'(x_1) = 0$, we must have $u(x) \equiv 0$. Hence λ is not an eigenvalue of A.

That the spectrum of A covers the whole λ-axis without gaps is shown in the same way as at the end of the proof of Theorem 5.5, and Theorem 5.19 is proved.

Note 5.20

We indicate that the condition $q(x) \in C^1(0, \infty)$ can be weakened. Theorem 5.19, for instance, remains true if we assume that $q(x)$ is only locally bounded. To show that this is so, Lemma 5.4 is proved once again under this weaker condition and for the special case $\sigma_\epsilon(x) = x + \epsilon$. If the difference (15) – (13) is multiplied by ϵ^{-1}, the result is

$$\int_0^\infty p'(x)[u'(x)]^2\varphi(x)\,dx +$$

$$\int_0^\infty (p'(x)u(x)u'(x) + p(x)u(x)u''(x) - p(x)[u'(x)]^2)\varphi'(x)\,dx + o(1) \tag{66}$$

$$= -\int_0^\infty [q(x+\epsilon) - q(x)]\epsilon^{-1}u(x+\epsilon)u(x)\varphi(x)\,dx\,,$$

where $o(1)$ refers to $\epsilon \to 0$. The integral on the right hand side of (66) is

$$\int_0^\infty [q(x+\epsilon) - q(x)]\epsilon^{-1}u^2(x)\varphi(x)\,dx$$

$$+\int_0^\infty [q(x+\epsilon) - q(x)]\epsilon^{-1}[u(x+\epsilon) - u(x)]u(x)\varphi(x)\,dx\,, \tag{67}$$

and we show that the second term here tends to zero when $\epsilon \to 0$. Using the mean value theorem of differential calculus and the fact that $u'(x)$ is uniformly continuous on the support of $\varphi(x)$, we obtain

$$\int_0^\infty [q(x+\epsilon) - q(x)]\epsilon^{-1}[u(x+\epsilon) - u(x)]u(x)\varphi(x)\,dx$$

$$= \int_0^\infty [q(x+\epsilon) - q(x)]u'(x)u(x)\varphi(x)\,dx + o(1) \tag{68}$$

as $\epsilon \to 0$. Again, since the factor $u'(x)u(x)\varphi(x)$ in the integral on the right-hand side is uniformly continuous on $[0.\ \infty)$, we obtain

$$\int_0^\infty [q(x+\epsilon) - q(x)]u'(x)u(x)\varphi(x)\,dx$$

$$= -\int_0^\infty q(x)u'(x)u(x)\varphi(x)\,dx + \int_0^\infty q(x+\epsilon)u'(x+\epsilon)u(x+\epsilon)\varphi(x+\epsilon)\,dx + o(1)$$

$$= o(1)\,. \tag{69}$$

as $\epsilon \to 0$. Hence, by (67), (68), and (69), we have

$$\int_0^\infty [q(x+\epsilon)-q(x)]\,\epsilon^{-1}u(x+\epsilon)u(x)\varphi(x)\,\mathrm{d}x$$

$$= \int_0^\infty [q(x+\epsilon)-q(x)]\,\epsilon^{-1}u^2(x)\varphi(x)\,\mathrm{d}x + o(1) \tag{70}$$

as $\epsilon \to 0$. If we substitute

$$p(x)u''(x) + p'(x)u'(x) = [q(x) - \lambda]\,u(x)$$

in (66), then from (66) and (70) we obtain

$$\int_0^\infty p'(x)[u'(x)]^2\varphi(x)\,\mathrm{d}x + \int_0^\infty ([q(x)-\lambda]\,u^2(x) - p(x)[u'(x)]^2)\varphi'(x)\,\mathrm{d}x$$

$$= -\int_0^\infty [q(x+\epsilon)-q(x)]\,\epsilon^{-1}u^2(x)\varphi(x)\,\mathrm{d}x + o(1)\,. \tag{71}$$

Since $q(x)$ is decreasing on (x_1, ∞), $[q(x+\epsilon)-q(x)]\,\epsilon^{-1}$ tends to $q'(x)$ almost everywhere as $\epsilon \to 0$, and then by the Fatou Lemma (see [48]) it follows from (71) that

$$-\int_0^\infty q'(x)u^2(x)\varphi(x)\,\mathrm{d}x$$

$$\leqslant \int_0^\infty p'(x)[u'(x)]^2\,\varphi(x)\,\mathrm{d}x + \int_0^\infty ([q(x)-\lambda]\,u^2(x) - p(x)[u'(x)]^2)\varphi'(x)\,\mathrm{d}x\,. \tag{72}$$

In (72) $\varphi(x)$ is taken to be the function $\varphi(x)$ from the proof of Theorem 5.19, where x_1 is chosen so large that

$$\lambda - q(x) > 0, \quad x_1 \leqslant x < \infty\,.$$

If $u(x)$ is not identically zero on (x_1, ∞), then the part

$$\int_{x_1}^{x_2} [q(x)-\lambda]\,u^2(x)\varphi'(x)\,\mathrm{d}x$$

of the right-hand side of (72) is negative. Then the right-hand side of (72) as a whole will be negative if we make $x_3 \to \infty$ and take (63) and (64) into account. On the other hand the left-hand side of (72) is non-negative, so that a contradiction arises. Hence we have $u(x) \equiv 0$ and λ is not an eigenvalue.

Generalizations of Theorem 5.19 for the case $p(x) = 1$ have been obtained by Eastham [12].

Remark 5.21

One can ask whether Theorem 5.19 remains true if $q(x)$ fails to satisfy (56). An example by Hartman and Wintner [16] for the special case $p(x) = 1$ shows, however, that for each $\eta > 0$ a decreasing function $q(x)$ with the properties

$$\int_0^\infty [q(x)]^{-1}\, dx > -\infty, \quad \int_0^\infty |q(x)|^{-1+\eta}\, dx = \infty$$

can be constructed so that the differential equation

$$-u'' + q(x)u = 0, \quad 0 < x < \infty,$$

has a non-trivial solution $u(x)$ for which

$$u(x) \in C^2[0, \infty) \cap L_2(0, \infty), \quad u(0) = 0\,.$$

This means that $\lambda = 0$ is an eigenvalue of a self-adjoint extension of A_0,

$$A_0 u = -u'' + q(x)u, \quad D(A_0) = C_0^\infty(0, \infty)\,.$$

Consequently, without (56) the condition $q'(x) \leqslant 0$, $x_1 < x$, is not sufficient for the continuity of the spectrum on $(-\infty, \infty)$. Conditions which do guarantee the continuity of the spectrum on $(-\infty, \infty)$ are known, and these involve $q'(x)$ and $q''(x)$ as well as $q(x)$. We refer to Hinton [19], where a more general differential expression is considered, and Walter [58]. For results concerning $\sigma_e(A) = (-\infty, \infty)$ without regard to eigenvalues, we refer to [54], [46], [15], [18].

Chapter 6

Oscillation Criteria

In this chapter we examine the question of under what conditions on the $a_k(x)$ is the differential equation

$$a[y] = \sum_{k=0}^{n} (-1)^k (a_k(x) y^{(k)})^{(k)} = 0\,,$$

$$a_k(x) \in W^k_{2,\mathrm{loc}}(0, \infty), \quad a_n(x) > 0\,, \quad 0 < x < \infty\,, \tag{1}$$

oscillatory or not at $x = \infty$ or $x = 0$ (see Definition 5.1). By Theorem 5.2, the question is equivalent to that of whether or not there exist an infinity of negative points in the spectrum of any self-adjoint extension of the operator A_0,

$$A_0 u = a[u]\,, \quad D(A_0) = C_0^\infty(0, \infty)\,.$$

Accordingly, we take up the problem of Theorem 2.19 once more. In the literature there are many results on the oscillation of (1), especially in the cases $n = 1$ and, for general n, $a_n(x) = 1$, $a_k(x) = 0$, $k = 1, \ldots, n-1$. Surveys of oscillation criteria can be found in the books by Glazman [15], Kreith [29] and Swanson [53]. The conditions in question on the coefficients are formulated as either integral con-conditions or pointwise conditions. The latter are also called "Kneser-type conditions", and we begin with these.

KNESER-TYPE CONDITIONS

The classical result of Kneser [27] is that the equation

$$-y'' + q(x)y = 0, \quad 0 < x < \infty, \quad q(x) \in C(0, \infty)\,, \tag{2}$$

is oscillatory or non-oscillatory at $x = \infty$ according as

$$\lim_{x\to\infty} x^2 q(x) < -\tfrac{1}{4} \quad \text{or} \quad \lim_{x\to\infty} x^2 q(x) > -\tfrac{1}{4}\,. \tag{3}$$

If the limit is equal to $-\frac{1}{4}$, then both cases are possible. This oscillation criterion of Kneser has several generalizations. Here we refer to the generalization of Hille [17] that (2) is oscillatory or not according as

$$\limsup_{x\to\infty} x^2\log^2 x\left(q(x)+\frac{1}{4x^2}\right)<-\tfrac{1}{4}$$

or (4)

$$\liminf_{x\to\infty} x^2\log^2 x\left(q(x)+\frac{1}{4x^2}\right)>-\tfrac{1}{4}\,.$$

In the following theorem we formulate conditions of the nature (4) for the equation

$$(-1)^n y^{(2n)}+q(x)y=0,\quad 0<x<\infty,\quad q(x)\in C(0,\infty),\quad n\geqslant 1\,, \tag{5}$$

and we use the notation $(2n-1)!!=(2n-1)(2n-3)(\cdots)3.1$.

Theorem 6.1
If there exists a point $x_0>0$ such that

$$q(x)>-\left(\sum_{j=0}^{n}\frac{[(2j-1)!!]^2\alpha_j}{4^j\log^{2j}x}\right)\frac{1}{x^{2n}},\quad x>x_0\,, \tag{6}$$

where the numbers α_j are the coefficients in the polynomial

$$\sum_{j=0}^{n}\alpha_j\xi^j=\left\{\xi+\left(\frac{1-2n}{2}\right)^2\right\}\left\{\xi+\left(\frac{5-2n}{2}\right)^2\right\}\cdots\left\{\xi+\left(\frac{2n-3}{2}\right)^2\right\},$$

then the equation (5) is non-oscillatory at $x=\infty$.

The equation (5) is oscillatory at $x=\infty$ if there exist a function $\delta(x)$ with the property

$$\int_a^\infty\frac{\delta(x)}{x\log x}\,dx=-\infty,\quad 1<a<\infty\ , \tag{7}$$

and a point $x_0>0$ such that

$$q(x)\leqslant-\left(\sum_{j=0}^{1}\frac{\alpha_j}{4^j\log^{2j}x}\right)\frac{1}{x^{2n}}+\frac{\delta(x)}{x^{2n}\log^2 x},\quad x>x_0\,. \tag{8}$$

Proof

First we show that (5) is non-oscillatory at $x = \infty$ under the condition (6). For this purpose the quadratic form

$$\int_0^\infty [(-1)^n u^{(2n)} + q(x)u]\bar{u}\,dx, \quad u(x) \in C_0^\infty(0, \infty),$$

will be estimated from below. The transformation

$$t = \log x, \quad u(x) = u(e^t) = P(t)e^{(n-1/2)t} \tag{9}$$

gives

$$u^{(2n)}(x) = \left(\sum_{l=0}^{2n} c_l P^{(l)}(t)\right) e^{-(n+1/2)t}, \quad c_{2n} = 1,$$

and we show now that the coefficients c_l have the properties

$$c_{2j-1} = 0, \quad c_{2j} = (-1)^{n-j}|c_{2j}|, \quad j = 0, \ldots, n.$$

The solutions of the differential equation

$$u^{(2n)}(x) = 0$$

are linear combinations of the powers x^ν, $\nu = 0, \ldots, 2n-1$. From the transformation (9) it follows that linearly independent solutions of the equation

$$\sum_{l=0}^{2n} c_l P^{(l)}(t) = 0 \tag{10}$$

are

$$\exp\{\pm(2k + \tfrac{1}{2} - n)t\}, \quad k = 0, \ldots, n-1. \tag{11}$$

The zeros of the characteristic polynomial of (10) are the constant factors in the exponents of the solutions (11), and so the characteristic polynomial is

$$\sum_{l=0}^{2n} c_l \lambda^l = \left\{\lambda^2 - \left(\frac{1-2n}{2}\right)^2\right\} \left\{\lambda^2 - \left(\frac{5-2n}{2}\right)^2\right\} \cdots \left\{\lambda^2 - \left(\frac{2n-3}{2}\right)^2\right\}. \tag{11A}$$

Hence we have

$$c_{2j-1} = 0, \quad c_{2j} = (-1)^{n-j} | c_{2j} |, \quad j = 1, \ldots, n,$$

and

$$c_0 = \frac{(-1)^n}{4^n} \prod_{\nu=0}^{n-1} (1 - 2n + 4\nu)^2 = \frac{(-1)^n}{4^n} [2n-1)!!]^2 . \tag{12}$$

Now the transformation (9) gives

$$\int_0^\infty (-1)^n u^{(2n)}(x) \overline{u(x)} \, \mathrm{d}x = \int_{-\infty}^\infty \left(\sum_{j=0}^n (-1)^j | c_{2j} | P^{(2j)}(t) \overline{P(t)} \right) \mathrm{d}t . \tag{13}$$

If we consider functions $u(x)$ whose supports lie in the interval $(1, \infty)$, then the corresponding functions $P(t)$ have compact support on the positive t-axis and, from (13), we have

$$\int_0^\infty (-1)^n u^{(2n)}(x) \overline{u(x)} \, \mathrm{d}x = \int_0^\infty \left(\sum_{j=0}^n | c_{2j} \| P^{(j)}(t) |^2 \right) \mathrm{d}t , \tag{14}$$

$$u(x) \in C_0^\infty (1, \infty) .$$

Using the estimate

$$\int_0^\infty | P^{(j)}(t) |^2 \, \mathrm{d}t \geqslant \frac{[(2j-1)!!]^2}{4^j} \int_0^\infty \frac{| P(t) |^2}{t^{2j}} \, \mathrm{d}t, \quad P(t) \in C_0^\infty (0, \infty) ,$$

which follows from the Hardy inequality (A.21) by iteration, we obtain from (14)

$$\int_0^\infty (-1)^n u^{(2n)}(x) \overline{u(x)} \, \mathrm{d}x \geqslant \int_0^\infty \left(\sum_{j=0}^n 4^{-j} t^{-2j} [(2j-1)!!]^2 | c_{2j} | \right) | P(t) |^2 \, \mathrm{d}t . \tag{15}$$

Defining

$$\rho_1(x) = - \left(\sum_{j=0}^n \frac{[(2j-1)!!]^2 \alpha_j}{4^j \log^{2j} x} \right) \frac{1}{x^{2n}} ,$$

we obtain from (15)

$$\int_0^\infty [(-1)^n u^{(2n)} + \rho_1(x)u]\bar{u}\,dx \geqslant$$

$$\int_0^\infty \left(\sum_{j=0}^{n} 4^{-j} t^{-2j} [(2j-1)!!]^2 (|c_{2j}| - \alpha_j)\right) |P(t)|^2\,dt\,.$$

By (11 A), $|c_{2j}| = \alpha_j$, and so we have

$$\int_0^\infty [(-1)^n u^{(2n)} + \rho_1(x)u]\bar{u}\,dx \geqslant 0, \quad u(x) \in C_0^\infty(1, \infty)\,. \tag{16}$$

If now $[\alpha, \beta]$ is an arbitrary interval with $\max(1, x_0) < \alpha < \beta < \infty$, then by (6) there is a number $\eta = \eta(\alpha, \beta) > 0$, such that

$$q(x) \geqslant \rho_1(x) + \eta, \quad x \in [\alpha, \beta]\,.$$

Hence from (16) we obtain

$$\int_\alpha^\beta [(-1)^n u^{(2n)} + q(x)u]\bar{u}\,dx \geqslant \eta \int_\alpha^\beta |u|^2\,dx, \eta > 0, \quad u(x) \in C_0^\infty(\alpha, \beta)\,.$$

Therefore there cannot be a nontrivial solution

$$y(x) \in \overset{\circ}{W}{}_2^n(\alpha, \beta) \cap W_{2,\mathrm{loc}}^{2n}(\alpha, \beta)$$

of (5), and the first part of Theorem 6.1 is proved.

To prove the second part, we construct a function $v(x) \in C_0^\infty(X, \infty)$, for a given $X > x_0$, such that

$$\int_X^\infty [(-1)^n v^{(2n)} + q(x)v]\bar{v}\,dx < 0\,. \tag{17}$$

We define

$$\rho_2(x) = -\left(\sum_{j=0}^{1} \frac{\alpha_j}{4^j \log^{2j} x}\right)\frac{1}{x^{2n}} + \frac{\delta(x)}{x^{2n}\log^2 x}\,,$$

and then, by (14) and (9), we obtain

$$\int_0^\infty [(-1)^n u^{(2n)} + \rho_2(x)u]\bar{u}\,\mathrm{d}x$$

$$= \int_0^\infty \left(\sum_{j=0}^{n} \alpha_j \mid P^{(j)}(t) \mid^2 - \sum_{j=0}^{1} 4^{-j} t^{-2j} \alpha_j \mid P(t) \mid^2 \right) \mathrm{d}t$$

$$+ \int_0^\infty t^{-2} \delta(e^t) \mid P(t) \mid^2 \mathrm{d}t \tag{18}$$

for $u(x) \in C_0^\infty(1, \infty)$. For $P(t)$ we choose

$$P(t) = e^{s/2} \varphi(s), \quad t = e^s, \quad -\infty < s < \infty,$$

where $\varphi(s)$ arises from

$$\chi(s) = \begin{cases} 1, & \sigma_1 \leqslant s < \sigma_2,\ 1 < \sigma_1 < \sigma_2 - 1, \\ \\ 0, & \text{otherwise} \end{cases}$$

by smoothing with radius $\frac{1}{2}$. Then, for certain constants a_{ji} and $b_{\mu\nu}$,

$$P^{(j)}(t) = e^{(1-2j)s/2} \,[a_{j0}\varphi(s) + a_{j1}\varphi'(s) + \cdots + a_{jj}\varphi^{(j)}(s)],$$

giving

$$\int_0^\infty \mid P^{(j)}(t) \mid^2 \mathrm{d}t = \int_0^\infty e^{2(1-j)s} \left(\sum_{0 \leqslant \mu, \nu \leqslant j} b_{\mu\nu} \varphi^{(\mu)}(s) \varphi^{(\nu)}(s) \right) \mathrm{d}s \to 0 \tag{19}$$

as $\sigma_1, \sigma_2 \to \infty$, provided that $j \geqslant 2$. Further we have

$$\int_0^\infty (\mid P'(t) \mid^2 - 4^{-1} t^{-2} \mid P(t) \mid^2)\, \mathrm{d}t = \int_0^\infty ([\varphi'(s)]^2 + \varphi(s)\varphi'(s))\, \mathrm{d}s = C < \infty \tag{20}$$

and

$$\int_0^\infty t^{-2} \delta(e^t) \mid P(t) \mid^2 \mathrm{d}t = \int_0^\infty \delta(\exp[e^s]) \varphi^2(s)\, \mathrm{d}s\,. \tag{21}$$

Using (19)–(21) in (18), we obtain

$$\int_0^\infty [(-1)^n u^{(2n)} + \rho_2(x)u]\bar{u}\,dx \leqslant C + \int_0^\infty \delta(\exp[e^s])\varphi^2(s)\,ds\,. \qquad (22)$$

Since the factor $\varphi^2(s)$ is increasing on $[\sigma_1 - \frac{1}{2}, \sigma_1 + \frac{1}{2}]$ from $\varphi^2(\sigma_1 - \frac{1}{2}) = 0$ to $\varphi^2(\sigma_1 + \frac{1}{2}) = 1$ and decreasing on $[\sigma_2 - \frac{1}{2}, \sigma_2 + \frac{1}{2}]$ from $\varphi^2(\sigma_2 - \frac{1}{2}) = 1$ to $\varphi^2(\sigma_2 + \frac{1}{2}) = 0$, a mean value theorem of integral calculus (see footnote p. 74) gives

$$\int_0^\infty \delta(\exp[e^s])\varphi^2(s)\,ds = \int_{\sigma_1'}^{\sigma_2'} \delta(\exp[e^s])\,ds = \int_{x_1'}^{x_2'} \frac{\delta(x)}{x\log x}\,dx\,,$$

$$\sigma_i - \tfrac{1}{2} \leqslant \sigma_i' \leqslant \sigma_i + \tfrac{1}{2}, \quad \sigma_i' = \log\log x_i', \quad i = 1, 2\,.$$

By (7) the integral on the right-hand side tends to $-\infty$ as $\sigma_2 \to \infty$ for fixed σ_1. Thus, by (22) and (8) we have the existence of a function $v(x) \in C_0^\infty(X, \infty)$ with the property (17). Then as in the proof of Theorem 5.2, we have an interval $[\alpha, \beta]$, $X < \alpha < \beta < \infty$, and a corresponding nontrivial solution

$$y(x) \in \overset{\circ}{W}{}_2^n(\alpha, \beta) \cap W_{2,\mathrm{loc}}^{2n}(\alpha, \beta)$$

of (5). Since X can be arbitrarily large the equation (5) is oscillatory at $x = \infty$, and Theorem 6.1 is proved.

Since the function $\delta(x) \equiv -\epsilon$ satisfies (7) for any $\epsilon > 0$ we can replace (8) by the stronger condition

$$\limsup_{x\to\infty} x^{2n} \log^2 x[q(x) + \alpha_0 x^{-2n}] < -\tfrac{1}{4}\alpha_1\,. \qquad (23)$$

We also note that, since $\alpha_0 = [(2n-1)!!]^2 4^{-n}$, the conditions (6) and (8) (or (23)) are less restrictive than those of Glazman [15], §30.

SPECIAL CASES

1) Let $n = 1$. Then the equation

$$-y'' + q(x)y = 0, \quad 0 < x < \infty\,,$$

is non-oscillatory or oscillatory at $x = \infty$ according as

$$q(x) > -\tfrac{1}{4}x^{-2}(1 + \log^{-2} x), \quad x > x_0\,,$$

or

$$q(x) \leqslant -\tfrac{1}{4}x^{-2}(1+\log^{-2}x) + \frac{\delta(x)}{x^2 \log^2 x}, \quad x > x_0,$$

$$\int_a^\infty \frac{\delta(x)}{x \log x}\, dx = -\infty, \quad 1 < a < \infty.$$

2) Let $n = 2$. The equation

$$u^{(4)} + q(x)u = 0, \quad 0 < x < \infty,$$

is non-oscillatory or oscillatory at $x = \infty$ according as

$$q(x) > -\tfrac{1}{16}x^{-4}(9 + 10\log^{-2}x + 9\log^{-4}x), \quad x > x_0,$$

or

$$q(x) \leqslant -\tfrac{1}{16}x^{-4}(9 + 10\log^{-2}x) + \frac{\delta(x)}{x^4 \log^2 x}, \quad x > x_0,$$

$$\int_a^\infty \frac{\delta(x)}{x \log x}\, dx = -\infty, \quad 1 < a < \infty.$$

These conditions weaken the well-known Kneser-type conditions of Leighton and Nehari [31] for this equation.

Concerning the oscillation of (5) at $x = 0$, one can prove the following theorem in the same way.

Theorem 6.2
If there exists a point $x_0 > 0$ such that

$$q(x) > -\left(\sum_{j=0}^{n} \frac{[(2j-1)!!]^2 \alpha_j}{4^j \log^{2j} x}\right) \frac{1}{x^{2n}}, \quad 0 < x < x_0,$$

where the α_j are as in Theorem 6.1, then equation (5) is non-oscillatory at $x = 0$.

Equation (5) is oscillatory at $x = 0$ if there exist a function $\delta(x)$ with the property

$$\int_0^a \frac{\delta(x)}{x\,|\log x|}\, dx = -\infty, \quad 0 < a < 1,$$

and a point $x_0 > 0$ such that

$$q(x) \leqslant -\left(\sum_{j=0}^{1} \frac{\alpha_j}{4^j \log^{2j} x}\right)\frac{1}{x^{2n}} + \frac{\delta(x)}{x^{2n}\log^2 x}, \quad 0<x<x_0 .$$

A direct consequence of Theorem 6.2 is

Theorem 6.3
Let

$$q(x) \in L_2(x_1, a) \text{ for every } x_1,\ 0<x_1<a,\ a<\infty,$$

and let

$$q(x) + \left(\sum_{j=0}^{n} \frac{[(2j-1)!!]^2\alpha_j}{4^j \log^{2j} x}\right)\frac{1}{x^{2n}} \to \infty \tag{24}$$

as $x \to 0$, where the α_j are as above. Then the spectrum of any self-adjoint extension of A_0,

$$A_0 u = (-1)^n u^{(2n)} + q(x)u,\quad D(A_0) = C_0^\infty(0, a),$$

is bounded from below and discrete.

Proof
If a constant $C>0$ is chosen arbitrarily, then by (24) there is a point $x_0>0$ such that

$$q(x) - C > -\left(\sum_{j=0}^{n} \frac{[(2j-1)!!]^2\alpha_j}{4^j \log^{2j} x}\right)\frac{1}{x^{2n}}, \quad 0<x<x_0 .$$

Then by Theorem 6.2 the differential equation

$$(-1)^n y^{(2n)} + [q(x) - C]y = 0,\quad 0<x<a,$$

is non-oscillatory at $x = 0$, which implies that the spectrum of any self-adjoint extension of $A_0 - CE$ possesses only a finite number of negative points. Since C can be arbitrarily large the spectrum of any self-adjoint extension of A_0 must be discrete.

In the case $n = 1$, Theorem 6.3 is a special case of a result of Berkowitz (see [8] and [51]).

INTEGRAL CONDITIONS

By a theorem of Leighton [30] the differential equation

$$-(p(x)y')' + q(x)y = 0, \ \ p(x) > 0, \ \ 0 < x < \infty,$$

is oscillatory at $x = \infty$ if the coefficients $p(x)$ and $q(x)$ satisfy the integral conditions

$$\int_1^\infty \frac{dx}{p(x)} = \infty \quad \text{and} \quad \int_1^\infty q(x)\,dx = -\infty,$$

In what follows this result is extended to the differential equation

$$(-1)^n (a_n(x)y^{(n)})^{(n)} + a_0(x)y = 0, \ \ a_n(x) > 0, \ \ 0 < x < \infty, \\ n \geqslant 1. \tag{25}$$

Theorem 6.4

Equation (25) is oscillatory at $x = \infty$ if

$$\int_1^\infty a_n^{-1}(x)\,dx = \infty \quad \text{and} \quad \int_1^\infty x^{2(n-1)} a_0(x)\,dx = -\infty.$$

Proof

We choose $\alpha (> 0)$ arbitrarily. The theorem is proved if we show that there is a point β, $\alpha < \beta < \infty$, and a nontrivial function $w(x) \in \mathring{W}_2^n(\alpha, \beta)$ such that the quadratic form

$$\int_\alpha^\beta a_n(x) \,|\, u^{(n)} \,|^2\,dx + \int_\alpha^\beta a_0(x) \,|\, u \,|^2\,dx, \ \ u(x) \in \mathring{W}_2^n(\alpha, \beta), \tag{26}$$

is negative for $u = w$. To construct $w(x)$ we need auxiliary functions $\omega_k(x)$, $k = 1, \ldots, n$, as follows. By smoothing of

$$\chi_\alpha(x) = \begin{cases} 0, & -\infty < x \leqslant \alpha + \frac{1}{2}, \\ 1, & \alpha + \frac{1}{2} < x < \infty, \end{cases}$$

with radius $\frac{1}{2}$, a function $\varphi_\alpha(x)$ arises with the properties

$$\varphi_\alpha(x) \begin{cases} = 0, & -\infty < x \leqslant \alpha, \\ = 1, & \alpha + 1 \leqslant x < \infty, \\ \in C^\infty(-\infty, \infty). \end{cases}$$

Then we define

$$\omega_k(x) = (\varphi_\alpha(x)x^n/n!)^{(n+1-k)}, \quad 0 \leqslant x \leqslant \alpha + 1, \quad 1 \leqslant k \leqslant n,$$

for which

$$\omega_k(\alpha + 1) = (\alpha + 1)^{k-1}/(k-1)!, \quad 1 \leqslant k \leqslant n.$$

In the construction of $w(x)$, we also need a further sequence $v_k(x)$, $k = 1, \ldots, n$, defined inductively. First we define $\alpha_1 = \alpha$ and

$$v_1(x) = \begin{cases} \omega_1(x), & 0 \leqslant x \leqslant \alpha_1 + 1, \\ 1, & \alpha_1 + 1 \leqslant x \leqslant \alpha_2, \\ 1 - \int_{\alpha_2}^{x} a_n^{-1}(t)\,\mathrm{d}t, & \alpha_2 \leqslant x \leqslant \beta_1, \text{ where } \int_{\alpha_2}^{\beta_1} a_n^{-1}(t)\,\mathrm{d}t = 1, \\ 0, & \beta_1 \leqslant x < \infty. \end{cases}$$

The point α_2 will be fixed later. Since

$$\int_{\alpha_2}^{\infty} a_n^{-1}(x)\,\mathrm{d}x = \infty$$

the point β_1 such that

$$\int_{\alpha_2}^{\beta_1} a_n^{-1}(x)\,\mathrm{d}x = 1$$

is uniquely determined by α_2. The restriction of $v_1(x)$ to (α_1, β_1) belongs to $\overset{\circ}{W}{}_2^1(\alpha_1, \beta_1)$. Then we define

$$v_{1,1}(x) = \begin{cases} v_1(x), & 0 \leqslant x \leqslant \beta_1, \\ 1 - \int_{\alpha_2}^{x} a_n^{-1}(t)\,\mathrm{d}t, & \beta_1 \leqslant x \leqslant \alpha_3, \quad \text{where } 1 - \int_{\alpha_2}^{\alpha_3} a_n^{-1}(t)\,\mathrm{d}t = \rho_1 \geqslant -1, \\ \rho_1, & \alpha_3 \leqslant x \leqslant \alpha_4, \; \alpha_3 \leqslant \alpha_4, \\ \rho_1 + \int_{\alpha_4}^{x} a_n^{-1}(t)\,\mathrm{d}t, & \alpha_4 \leqslant x \leqslant \beta_2, \quad \text{where } \rho_1 + \int_{\alpha_4}^{\beta_2} a_n^{-1}(t)\,\mathrm{d}t = 0, \\ 0, & \beta_2 \leqslant x < \infty. \end{cases}$$

Here, the points α_3, α_4 and β_2, and the value of ρ_1, are uniquely determined by α_1 and α_2 if we also require that

$$\int_{\alpha_1}^{\beta_1} v_{1,1}(x)\,dx = -\int_{\beta_1}^{\beta_2} v_{1,1}(x)\,dx .$$

We note that the following cases can occur:

1) $\alpha_3 = \alpha_4, -1 < \rho_1 < 0$

2) $\alpha_3 = \alpha_4,\ \rho_1 = -1$

3) $\alpha_3 < \alpha_4,\ \rho_1 = -1$.

We say that $v_{1,1}(x)$ is obtained from $v_1(x)$ by "symmetric continuation of $v_1(x)$ at the point β_1",† and we say that the pairs of points (α_1, β_2), $(\alpha_1 + 1, \alpha_4)$, (α_2, α_3) correspond in the symmetric continuation.

Continuing with the construction of $w(x)$, we can now define

$$v_2(x) = \int_0^x v_{1,1}(t)\,dt, \quad 0 < x < \infty .$$

Then $v_2(x)$ has the following properties.

1) $v_2(x) \in \mathring{W}_2^2(\alpha_1, \beta_1)$,

2) $v_2(x) = 0,\ x \in [0, \alpha_1] \cup [\beta_2, \infty)$,

3) $v_2(x) = \omega_2(x),\ 0 \leqslant x \leqslant \alpha_1 + 1$,

4) $v_2'(x) > 0,\ \alpha_1 + 1 \leqslant x < \beta_1;\ v_2'(x) < 0,\ \beta_1 < x < \beta_2$,

5) $v_2(x) = x f_2(x);\ f_2(x) = 1,\ \alpha_1 + 1 \leqslant x \leqslant \alpha_2;\ f_2(x) > 0,\ f_2'(x) < 0,\ \alpha_2 < x < \beta_2$

(1)–(4) are obvious. To prove (5) we start with the inequality

$$x v'(x) = -x/a_n(x) < 0, \quad \alpha_2 < x \leqslant \beta_1 ,$$

and, since $v_1(\alpha_2) = 1$ and $v_2(\alpha_2) = \alpha_2$, we obtain

$$0 > \int_{\alpha_2}^{x} t v_1'(t)\,dt = x v_1(x) - \alpha_2 v_1(\alpha_2) - \int_{\alpha_2}^{x} v_1(t)\,dt = x v_2'(x) - v_2(x)$$

†Editor's note: A sketch of the graph of $v_{1,1}(x)$ will help to visualize the situation.

in $(\alpha_2, \beta_1]$. It follows that $f_2'(x) < 0$ in this interval. Since also $f_2'(x) < 0$ in (β_1, β_2) by Property (4), this proves Property (5).

To define $v_3(x)$, we first return to $v_{1,1}(x)$ and we symmetrically continue $v_{1,1}(x)$ at β_2 to obtain a new function $v_{1,2}(x)$. The new points introduced to the right of β_2 correspond to the existing points as follows: (α_1, β_4), $(\alpha_1 + 1, \alpha_8)$, (α_2, α_7), (β_1, β_3), (α_3, α_6), (α_4, α_5), and they are uniquely determined if we also require that

$$\int_{\beta_2}^{\beta_3} v_{1,2}(x)\,dx = -\int_{\beta_3}^{\beta_4} v_{1,2}(x)\,dx \ (< 0)$$

and

$$\int_{\beta_2}^{\beta_4} v_{2,2}(x)\,dx = -\int_{\alpha_1}^{\beta_2} v_2(x)\,dx \ (< 0),$$

where

$$v_{2,2}(x) = \int_0^x v_{1,2}(x)\,dt$$

We note that $v_{2,2}(x) = v_2(x)$, $0 \leqslant x \leqslant \beta_2$. Now we define

$$v_3(x) = \int_0^x v_{2,2}(t)\,dt = \int_0^x \int_0^{t_2} v_{1,2}(t_1)\,dt_1\,dt_2, \quad 0 \leqslant x < \infty.$$

Finally, continuing this method, we obtain a function

$$v_n(x) = \int_0^x \int_0^{t_{n-1}} \cdots \int_0^{t_2} v_{1,n-1}(t_1)\,dt_1 \ldots dt_{n-1}, \quad 0 \leqslant x < \infty,$$

with the properties

1) $v_n(x) \in \mathring{W}_2^n(\alpha_1, \beta(n))$,

2) $v_n(x) = 0$, $x \in [0, \alpha_1] \cup [\beta(n), \infty)$, $v_n(x) > 0$, $\alpha_1 < x < \beta(n)$,

3) $v_n(x) = \omega_n(x)$, $0 \leqslant x \leqslant \alpha_1 + 1$,

4) $v_n'(x) > 0$, $\alpha_1 + 1 \leqslant x < \beta(n-1)$; $v_n'(x) < 0$, $\beta(n-1) < x < \beta(n)$,

5) $v_n(x) = \dfrac{x^{n-1}}{(n-1)!} f_n(x)$; $f_n(x) = 1$, $\alpha_1 + 1 \leqslant x \leqslant \alpha_2$; $f_n'(x) < 0$, $\alpha_2 < x < \beta(n)$,

where $\beta(n)$ denotes β_k with $k = 2^{n-1}$. These properties can be proved by induction and we do this for (5). The result holds for $n = 2$ as we noted above. If we now assume that

$$v_{n-1}(x) = \frac{x^{n-2}}{(n-2)!} f_{n-1}(x) \quad \text{where} \quad f'_{n-1}(x) < 0, \quad \alpha_2 < x < \beta(n-1),$$

then $(v_{n-1}(x)/x^{n-2})' < 0$ and hence

$$x v'_{n-1} - (n-2) v_{n-1} < 0, \quad \alpha_2 < x < \beta(n-1),$$

Integrating and noting that

$$v_k(\alpha_2) = \frac{\alpha_2^{k-1}}{(k-1)!}, \quad 1 \leqslant k \leqslant n,$$

we obtain

$$0 > \int_{\alpha_2}^{x} t v'_{n-1}\, dt - (n-2) \int_{\alpha_2}^{x} v_{n-1}\, dt$$

$$= x v_{n-1}(x) - \alpha_2 v_{n-1}(\alpha_2) - v_n(x) + v_n(\alpha_2) - (n-2) v_n(x) + (n-2) v_n(\alpha_2)$$

$$= x v_{n-1}(x) - (n-1) v_n(x) = x v'_n(x) - (n-1) v_n(x)$$

for $\alpha_2 < x < \beta(n-1)$. Thus, writing $v_n(x) = x^{n-1} f_n(x)/(n-1)!$, we have

$$f'_n(x) < 0, \quad \alpha_2 < x < \beta(n-1).$$

Since we also have $f'_n(x) < 0$ for $\beta(n-1) \leqslant x < \beta(n)$ by Property (4), Property (5) follows immediately.

The function $v_n(x)$ is uniquely determined by the points α_1 and α_2 and we take $v_n(x)$ to be the function $w(x)$ that we are constructing. We calculate the quadratic form (26) for $u(x) = w(x)$ where $\beta = \beta(n)$. Writing $N = 2^n$, we have

$$\int_{\alpha_1}^{\beta(n)} a_n(x)[w^{(n)}(x)]^2\, dx + \int_{\alpha_1}^{\beta(n)} a_0(x) w^2(x)\, dx$$

$$= \int_{\alpha_1}^{\alpha_1+1} a_n(x)[\omega'_1(x)]^2\, dx + \int_{\alpha_2}^{\alpha_3} a_n(x) a_n^{-2}(x)\, dx + \cdots + \int_{\alpha_{N-2}}^{\alpha_{N-1}} a_n(x) a_n^{-2}(x)\, dx \tag{27}$$

$$+ \int_{\alpha_N}^{\beta(n)} a_n(x) a_n^{-2}(x)\, dx + \int_{\alpha_1}^{\alpha_1+1} a_0(x)\omega_n^2(x)\, dx + \frac{1}{[(n-1)!]^2} \int_{\alpha_1+1}^{\beta(n)} a_0(x) x^{2n-2} f_n^2(x)\, d$$

Since $f_n^2(x)$ is decreasing on $[\alpha_2, \beta(n)]$ from $f_n^2(\alpha_2) = 1$ to $f_n^2(\beta(n)) = 0$ there exists a point $\xi_n, \alpha_2 \leqslant \xi_n \leqslant \beta(n)$, such that

$$\int_{\alpha_1+1}^{\beta(n)} a_0(x)x^{2n-2} f_n^2(x)\,\mathrm{d}x = \int_{\alpha_1+1}^{\xi_n} a_0(x)x^{2n-2}\,\mathrm{d}x\,.$$

The middle terms on the right-hand side of (27) are estimated from

$$\int_{\alpha_{2\nu}}^{\alpha_{2\nu+1}} a_n^{-1}(x)\,\mathrm{d}x \leqslant 2, \quad \nu = 1, \ldots, 2^{n-1}-1, \text{ and } \int_{\alpha_N}^{\beta(n)} a_n^{-1}(x)\,\mathrm{d}x \leqslant 1\,.$$

This gives

$$\int_{\alpha_1}^{\beta(n)} a_n(x)[w^{(n)}(x)]^2\,\mathrm{d}x + \int_{\alpha_1}^{\beta(n)} a_0(x)w^2(x)\,\mathrm{d}x$$

$$\leqslant \int_{\alpha_1}^{\alpha_1+1} a_n(x)[\omega_1'(x)]^2\,\mathrm{d}x + 2^n - 1 + \int_{\alpha_1}^{\alpha_1+1} a_0(x)\omega_n^2(x)\,\mathrm{d}x$$

$$+ \frac{1}{[(n-1)!]^2} \int_{\alpha_1+1}^{\xi_n} a_0(x)x^{2n-2}\,\mathrm{d}x, \quad \alpha_2 \leqslant \xi_n\,.$$

By hypothesis the last term on the right-hand side of this inequality tends to $-\infty$ when $\alpha_2 \to \infty$. Hence if $\alpha = \alpha_1$ is fixed the value of the quadratic form is negative when α_2 is chosen sufficiently large. This proves Theorem 6.4.

Under the restriction $a_0(x) \leqslant 0$ Theorem 6.4 was proved by Hunt and Namboordiri [22].

Next we formulate oscillation criteria for the equation (1), again in the neighbourhood of $x = \infty$.

Theorem 6.5
For some constant σ, let

$$\int_1^\infty x^{\sigma-2k} a_k^+(x)\,\mathrm{d}x < \infty, \quad a_k^+(x) = \max(a_k(x), 0), \quad 1 \leqslant k \leqslant n\,, \tag{28}$$

and

$$\int_1^\infty x^\sigma a_0(x)\,\mathrm{d}x = -\infty\,.$$

Then equation (1) is oscillatory at $x = \infty$.

Proof
We show that, for any given $\alpha > 0$, there is a point $\beta > \alpha$ and a nontrivial function $v(x) \in C_0^\infty(\alpha, \beta)$ such that the quadratic form $(a[u], u)_{(\alpha,\beta)}$, $u \in C_0^\infty(\alpha, \beta)$, takes a negative value for $u = v$. By smoothing the functions

$$\chi_1(x) = \begin{cases} 0, \ x \leqslant \alpha + 1 , \\ 1, \ x > \alpha + 1 , \end{cases}$$

and

$$\chi_2(x) = \begin{cases} 1, x \leqslant \frac{3}{2} , \\ 0, \ x > \frac{3}{2} , \end{cases}$$

with radius $\frac{1}{2}$, we obtain functions $\varphi_1(x)$ and $\varphi_2(x)$ which we use to define the test function

$$v(x) = \varphi_1(x) x^{\sigma/2} \ \varphi_2\left(\frac{x}{R}\right), \quad R > \alpha + 1, \quad \beta > 2R .$$

Then we have

$$\begin{aligned}(a[v], v)_{(\alpha,\beta)} \leqslant & \sum_{k=1}^{n} \int_\alpha^\beta a_k^+(x)[v^{(k)}(x)]^2 \, \mathrm{d}x + \int_\alpha^\beta a_0(x) v^2(x) \, \mathrm{d}x \\ & + \sum_{k=1}^{n} \int_{\alpha+1/2}^{\alpha+3/2} a_k(x)[v^{(k)}]^2 \, \mathrm{d}x + \sum_{k=1}^{n} \int_{\alpha+3/2}^{R} c_{\sigma,k} a_k^+(x) x^{\sigma-2k} \, \mathrm{d}x \\ & + \sum_{k=1}^{n} \int_R^{2R} a_k^+(x) \left(\left\{ x^{\sigma/2} \varphi_2\left(\frac{x}{R}\right) \right\}^{(k)} \right)^2 \mathrm{d}x \\ & + \int_{\alpha+1/2}^{2R} x^\sigma a_0(x) \varphi_1^2(x) \varphi_2^2\left(\frac{x}{R}\right) \mathrm{d}x . \end{aligned} \tag{30}$$

The third sum on the right-hand side of (30) has the form

$$\sum_{k=1}^{n} \int_R^{2R} a_k^+(x) x^{\sigma-2k} b_k(x) \, \mathrm{d}x ,$$

where the $b_k(x)$ are bounded on $[R, 2R]$ uniformly in R. The last term in (30) is dealt with by the mean value theorem of integral calculus. Thus, there exist numbers ξ_1, $\alpha + \frac{1}{2} \leqslant \xi_1 \leqslant \alpha + \frac{3}{2}$, and ξ_2, $R \leqslant \xi_2 \leqslant 2R$, such that

$$\int_{\alpha+1/2}^{2R} x^{\sigma} a_0(x)\varphi_1^2(x)\varphi_1^2\left(\frac{x}{R}\right) \mathrm{d}x = \int_{\xi_1}^{\xi_2} x^{\sigma} a_0(x)\, \mathrm{d}x\, .$$

The first sum on the right-hand side of (30) depends only on α and we write it as S_α. Then from (30) we obtain

$$(a[v],v)_{(\alpha,\beta)} \leqslant S_\alpha + C \sum_{k=1}^{n} \int_{\alpha+3/2}^{2R} x^{\sigma-2k} a_k^+(x)\, \mathrm{d}x + \int_{\xi_1}^{\xi_2} x^{\sigma} a_0(x)\, \mathrm{d}x\, . \tag{31}$$

Now, by (28) and (29), the right-hand side of (31) is negative if R is chosen sufficiently large, and so Theorem 6.5 is proved.

In the case $\sigma = 0$ the second sum on the right-hand side of (30) is absent and then (31) becomes

$$\begin{aligned}(a[v],v)_{(\alpha,\beta)} &\leqslant S_\alpha + C \sum_{k=1}^{n} \int_{R}^{2R} x^{-2k} a_k^+(x)\, \mathrm{d}x + \int_{\xi_1}^{\xi_2} a_0(x)\, \mathrm{d}x \\ &= S'_\alpha + C \sum_{k=1}^{n} \int_{R}^{2R} x^{-2k} a_k^+(x)\, \mathrm{d}x + \int_{1}^{\xi_2} a_0(x)\, \mathrm{d}x\, .\end{aligned} \tag{32}$$

Restricting the coefficients by

$$a_k^+(x) \leqslant C_k x^{\alpha_k}, \quad x > 1, \quad 1 \leqslant k \leqslant n\, , \tag{33}$$

and writing

$$\gamma = \max_{1\leqslant k\leqslant n} \{\alpha_k - 2k\}\, , \tag{34}$$

from (32) we obtain

$$\begin{aligned}(a[v],v)_{(\alpha,\beta)} &\leqslant C_\alpha + C\xi_2^{\gamma+1} + \int_{1}^{\xi_2} a_0(x)\, \mathrm{d}x \\ &= C_\alpha + \xi_2^{\gamma+1}\left(C + \xi_2^{-(\gamma+1)} \int_{1}^{\xi_2} a_0(x)\, \mathrm{d}x\right),\end{aligned} \tag{35}$$

provided that $\gamma \geqslant -1$. Thus we have

Corollary 6.6

Let the coefficients of equation (1) satisfy (33) and, with γ defined by (34), let $\gamma \geqslant -1$. Then (1) is oscillatory at $x = \infty$ if

$$x^{-(1+\gamma)} \int_1^x a_0(t)\,\mathrm{d}t \to -\infty \quad \text{as} \quad x \to \infty .$$

In particular, if

$$a_k^+(x) \leqslant C_k x^{2k-1}, \quad x > 1, \quad 1 \leqslant k \leqslant n ,$$

then (1) is oscillatory at $x = \infty$ if also

$$\int_1^\infty a_0(x)\,\mathrm{d}x = -\infty .$$

Under somewhat stronger assumptions on the $a_k(x)$ Theorem 6.5 was proved by Lewis [34]. Also in [34] it is shown that (1) is non-oscillatory at $x = \infty$ if $a_n(x) \equiv 1$ and

$$\int_1^\infty x^{2(n-k)-1} \,|\, a_k(x) \,|\, \mathrm{d}x < \infty, \quad 0 \leqslant k \leqslant n-1, \quad (\sigma = 2n-1) .$$

Concerning equation (25), oscillation criteria have been given by Lewis [33] which are similar to the one in Corollary 6.6.

The use of other test functions yields new oscillation theorems such as the following one.

Theorem 6.7

For some $\sigma \geqslant 0$ and $\mu > 1$ let

$$\limsup_{x \to \infty} x^{\sigma-2k} \int_x^{\mu x} a_k^+(t)\,\mathrm{d}t < \infty, \quad a_k^+(x) = \max(a_k(x), 0), \quad 1 \leqslant k \leqslant n , \tag{36}$$

and

$$\lim_{x \to \infty} x^{\sigma} \int_x^\infty a_0(t)\,\mathrm{d}t = -\infty . \tag{37}$$

Then (1) is oscillatory at $x = \infty$. If further $a_0(x) \leqslant 0$, $x > x_0$, then (37) can be replaced by

$$\liminf_{x\to\infty} x^\sigma \int_x^\infty a_0(t)\,dt = -\infty. \tag{38}$$

The particular equation

$$(-1)^n(a_n(x)u^{(n)})^{(n)} + a_0(x)u = 0, \quad a_n(x) > 0, \quad 0 < x < \infty, \tag{39}$$

is oscillatory at ∞ if (37) holds and (36) is weakened to

$$\liminf_{x\to\infty} x^{\sigma-2n} \int_x^{\mu x} a_n(t)\,dt < \infty \tag{40}$$

for some $\sigma \geqslant 0$ and $\mu > 1$.

Proof
By smoothing of

$$\chi(x) = \begin{cases} 0, & -\infty < x \leqslant \frac{1}{2}(1+\mu), \\ 1, & \frac{1}{2}(1+\mu) < x < \infty, \end{cases}$$

with radius $\frac{1}{2}(\mu - 1)$, we obtain a function $\varphi_1(x)$ with the properties

$$\varphi_1(x) \begin{cases} = 0, \ -\infty < x \leqslant 1, \\ = 1, \ \mu \leqslant x < \infty, \\ \in C^\infty(-\infty, \infty), \ \varphi'(x) > 0, \ 1 < x < \mu. \end{cases}$$

Then we define the functions

$$\varphi_R(x) = \varphi_1\left(\frac{x}{R}\right), \ R \geqslant 1, \quad \text{and} \quad \psi_T(x) = 1 - \varphi_T(x), \quad T > \mu R, \quad -\infty < x < \infty,$$

which we use to define the test function

$$v(x) = R^{\sigma/2}\varphi_R(x)\psi_T(x), \quad 0 < x < \infty.$$

Then we have

$$(\mathfrak{a}[v], v)_{(R,\mu T)} = \sum_{k=0}^{n} \int_R^{\mu T} a_k(x)[v^{(k)}(x)]^2\,dx$$

$$\leqslant \sum_{k=1}^{n} \int_R^{\mu T} a_k^+(x)[v^{(k)}(x)]^2\,dx + \int_R^{\mu T} a_0(x)v^2(x)\,dx$$

$$\leqslant \sum_{k=1}^{n} R^{\sigma-2k} \int_R^{\mu R} b_k(x)a_k^+(x)\,dx + \sum_{k=1}^{n} R^{\sigma}T^{-2k} \int_T^{\mu T} b_k(x)a_k^+(x)\,dx$$

$$+ R^{\sigma} \int_R^{\mu T} a_0(x)\varphi_R^2\,\psi_T^2\,dx\,,$$

where the $b_k(x)$ are bounded and do not depend on R and T. Hence we obtain

$$(a[v], v)_{(R,\mu T)} \leqslant C \sum_{k=1}^{n} R^{\sigma-2k} \int_R^{\mu R} a_k^+(x)\,dx + C \sum_{k=1}^{n} R^{\sigma}T^{-2k} \int_T^{\mu T} a_k^+(x)\,dx$$

$$+ R^{\sigma} \int_R^{\mu T} a_0(x)\varphi_R^2\,\psi_T^2\,dx\,.$$

Since $\sigma \geqslant 0$, the factor R^{σ} in the second sum on the right-hand side can be replaced by T^{σ}. Then from (36) it follows that the first two sums are bounded by a constant independent of R and T. Hence

$$(a[v], v)_{(R,\mu T)} \leqslant C + R^{\sigma} \int_R^{\mu T} a_0(x)\varphi_R^2\,\psi_T^2\,dx\,. \tag{41}$$

First we consider the case that $a_0(x)$ has no fixed sign for large x. Then there exist numbers $\xi_1, R \leqslant \xi_1 \leqslant \mu R$, and $\xi_2, T \leqslant \xi_2 \leqslant \mu T$, such that

$$R^{\sigma} \int_R^{\mu T} a_0(x)\varphi_R^2\,\psi_T^2\,dx = R^{\sigma} \int_{\xi_1}^{\xi_2} a_0(x)\,dx\,. \tag{42}$$

By (37), R can be chosen so large ($R \geqslant R_0$) that

$$\int_{\xi_1}^{\xi_2} a_0(x)\,dx < 0$$

for all $T \geqslant T_0(R)$. For such parameters R and T we have

$$R^{\sigma} \int_{\xi_1}^{\xi_2} a_0(x)\,dx \leqslant \mu^{-\sigma}\xi_1^{\sigma} \int_{\xi_1}^{\xi_2} a_0(x)\,dx\,.$$

Hence by (41) and (42) we obtain

$$(a[v], v)_{(R,\mu T)} \leqslant C + \mu^{-\sigma}\xi_1^{\sigma}\int_{\xi_1}^{\xi_2} a_0(x)\,dx\,, \tag{43}$$

from which we see that R and a corresponding T can be chosen so large that

$$(a[v], v)_{(R,\mu T)} < 0\,. \tag{44}$$

In the case that $a_0(x) \leqslant 0, x > x_0$, from (41) we have

$$(a[v], v)_{(R,\mu T)} \leqslant C + \mu^{-\sigma}(\mu R)^{\sigma}\int_{\mu R}^{T} a_0(x)\,dx, \quad \mu R > x_0\,.$$

Then, by (38), it follows that for selected parameters R and T (44) holds again, and hence that (1) is oscillatory at $x = \infty$.

The weakening of (36) to (40) in the case of (39) is easily checked and we omit these details.

SPECIAL CASES

1) $\sigma = 0$. In the case $a_n(x) \equiv 1$, $a_k(x) \equiv 0$, $1 \leqslant k \leqslant n-1$, we obtain results of Wintner [60] ($n = 1$) and Glazman [15] ($n \geqslant 1$).

2) $\sigma = 2n - 1$. Equation (39) is oscillatory at $x = \infty$ if

$$\limsup_{x\to\infty} x^{-1}\int_x^{\mu x} a_n(t)\,dt < \infty \tag{45}$$

and either

$$\lim_{x\to\infty} x^{2n-1}\int_x^{\infty} a_0(t)\,dt = -\infty\,, \tag{46}$$

or

$$\liminf_{x\to\infty} x^{2n-1}\int_x^{\infty} a_0(t)\,dt = -\infty \quad \text{and} \quad a_0(x) \leqslant 0, \quad x > x_0\,. \tag{47}$$

Certainly (45) is satisfied if $a_n(x) \equiv 1$. From the proof of Theorem 6.7 it can be seen that (47) can be weakened in the following sense for the equation

$$(-1)^n u^{(2n)} + a_0(x)u = 0, \quad 0 < x < \infty, \quad a_0(x) \leqslant 0, \quad x > x_0\,.$$

There exists a constant C^*, $0 < C^* < \infty$, such that

$$\liminf_{x\to\infty} x^{2n-1} \int_x^\infty a_0(t)\,dt < -C^* .$$

Glazman [15] calculated the constant

$$\left\{ \frac{\sqrt{(2n-1)}}{(n-1)!} \sum_{k=1}^{n} (-1)^{k-1}(2n-k)^{-1} \binom{n-1}{k-1} \right\}^{-2}$$

as a possible C^*. See also [21] and [33]. In the case $n = 1$ this constant is equal to 1 and is the best possible one, as shown by Hille [17].

THE STURM-LIOUVILLE EQUATION

A well-known result of Nehari [45] states that the equation

$$-y'' + q(x)y = 0,\quad q(x) \leqslant 0,\quad 0 < x < \infty ,$$

is oscillatory at $x = \infty$ if

$$\liminf_{x\to\infty} x^{1-\sigma} \int_x^\infty t^\sigma q(t)\,dt < -\tfrac{1}{4}(2-\sigma)^2/(1-\sigma),\quad 0 \leqslant \sigma < 1 . \qquad (48)$$

In the case $\sigma = 0$, this condition reduces to the condition of Hille [17] which we mentioned above. The oscillation criteria which we next obtain for the equation

$$a[y] \equiv -(x^\rho y')' + q(x)y = 0,\quad 0 < x < \infty,\quad -\infty < \rho < \infty , \qquad (49)$$

are of the nature (48), and the condition $q(x) \leqslant 0$ is not required.

Theorem 6.8
The equation (49) is oscillatory at $x = \infty$ if

$$\limsup_{x\to\infty} x^{1-\rho-\sigma} \int_x^\infty t^\sigma q(t)\,dt < \tfrac{1}{2}(\rho - 1 - |\rho - 1|) + \tfrac{1}{4}\sigma^2/(\rho+\sigma-1),$$
$$\rho + \sigma < 1 , \qquad (50)$$

or

$$\limsup_{x\to\infty} x^{1-\rho-\sigma} \int_1^x t^\sigma q(t)\,dt < \tfrac{1}{2}(1 - \rho - |\rho - 1|) - \tfrac{1}{4}\sigma^2/(\rho+\sigma-1),$$
$$\rho + \sigma > 1 . \qquad (51)$$

In the case $q(x) \leqslant 0, x > x_0$, the assertions are also true if "lim sup" is replaced by "lim inf".

In the case where $\rho + \sigma = 1$, (50) and (51) are replaced by

$$\limsup_{x\to\infty} \frac{1}{\log x} \int_1^x t^{1-\rho} q(t)\,dt < -\frac{(1-\rho)^2}{4}, \tag{52}$$

and again "lim sup" can be replaced by "lim inf" if $q(x) \leqslant 0, x > x_0$.

Proof

Again we show that for any given $\alpha > 0$ there exist a point $\beta > \alpha$ and a function $v(x) \in \overset{\circ}{W}{}_2^1(\alpha, \beta)$ such that

$$(a[v], v)_{(\alpha,\beta)} < 0 .$$

First let us assume that $\rho \neq 1$. We introduce parameters R and T and define

$$v(x) = \begin{cases} R^{\sigma/2}(x^{1-\rho} - \alpha^{1-\rho})/(R^{1-\rho} - \alpha^{1-\rho}), & \alpha \leqslant x \leqslant R, \\ x^{\sigma/2}, & R \leqslant x \leqslant T, \\ T^{\sigma/2}(x^{1-\rho} - \beta^{1-\rho})/(T^{1-\rho} - \beta^{1-\rho}), & T \leqslant x \leqslant \beta, \end{cases} \tag{53}$$

where, on the intervals $[\alpha, R]$ and $[T, \beta]$, the function of $v(x)$ is chose in such a way that the terms

$$\int_\alpha^R x^\rho (v')^2\,dx \quad \text{and} \quad \int_T^\beta x^\rho (v')^2\,dx$$

in the quadratic form $(a[v], v)_{(\alpha,\beta)}$ are minimal. †

† The Euler differential equation in the calculus of variations,

$$F_v - \frac{d}{dx} F_{v'} \equiv -2\rho x^{\rho-1} v' - 2x^\rho v'' = 0, \quad v(\alpha) = 0, \quad v(R) = R^{\sigma/2},$$

is a necessary condition for $v(x)$ to give the functional

$$\int_\alpha^R F(x, v, v')\,dx \equiv \int_\alpha^R x^\rho (v')^2\,dx, \quad v(\alpha) = 0, \quad v(R) = R^{\sigma/2},$$

a minimal value. The function $v(x)$, $\alpha \leqslant x \leqslant R$, defined in (53) is the solution of the Euler differential equation.

We now discuss the case $\rho < 1, \rho + \sigma < 1$. We have

$$(a[v], v)_{(\alpha,\beta)} = R^{\rho+\sigma-1}\left(\frac{1-\rho}{1-(\alpha/R)^{1-\rho}} + \frac{\sigma^2}{4(\rho+\sigma-1)}\left\{\left(\frac{T}{R}\right)^{\rho+\sigma-1} - 1\right\}\right. \tag{54}$$

$$\left. - \left(\frac{T}{R}\right)^{\rho+\sigma-1} \frac{1-\rho}{1-(\beta/R)^{1-\rho}} + R^{1-\rho-\sigma}\int_\alpha^\beta q(x)x^\sigma \varphi^2(x)\,\mathrm{d}x\right)$$

where

$$\varphi^2(x) = \begin{cases} x^{-\sigma}R^\sigma(x^{1-\rho} - \alpha^{1-\rho})^2/(R^{1-\rho} - \alpha^{1-\rho})^2, & \alpha \leqslant x \leqslant R, \\ 1, R \leqslant x \leqslant T, \\ x^{-\sigma}T^\sigma(x^{1-\rho} - \beta^{1-\rho})^2/(T^{1-\rho} - \beta^{1-\rho})^2, & T \leqslant x \leqslant \beta. \end{cases}$$

A short calculation shows that $\varphi^2(x)$ is increasing on $[\alpha, R]$ for any α and R, $\alpha < R$, and decreasing on $[T, \beta]$ if T is sufficiently close to β. Therefore there exist numbers α_1, $\alpha \leqslant \alpha_1 \leqslant R$, and β_1, $T \leqslant \beta_1 \leqslant \beta$, such that

$$R^{1-\rho-\sigma}\int_\alpha^\beta q(x)x^\sigma\varphi^2(x)\,\mathrm{d}x = R^{1-\rho-\sigma}\int_{\alpha_1}^{\beta_1} x^\sigma q(x)\,\mathrm{d}x\,. \tag{55}$$

By (50), we can choose the parameters α, R and T so that

$$R^{1-\rho-\sigma}\int_{\alpha_1}^{\beta_1} x^\sigma q(x)\,\mathrm{d}x \leqslant \alpha_1^{1-\rho-\sigma}\int_{\alpha_1}^{\beta_1} x^\sigma q(x)\,\mathrm{d}x < \rho - 1 + \frac{\sigma^2}{4(\rho+\sigma-1)} - 3\delta\,, \tag{56}$$

where $\alpha > \alpha_\delta$, $T > T_R$, and $\delta > 0$ is sufficiently small.

In the first term in the bracket on the right-hand side of (54), the parameter R can be chosen so large that

$$\frac{1-\rho}{1-(\alpha/R)^{1-\rho}} < 1 - \rho + \delta, \quad R > R_{\alpha,\delta}\,. \tag{57}$$

Further we have

$$\frac{\sigma^2}{4(\rho+\sigma-1)}\left\{\left(\frac{T}{R}\right)^{\rho+\sigma-1} - 1\right\} - \left(\frac{T}{R}\right)^{\rho+\sigma-1}\frac{1-\rho}{1-(\beta/T)^{1-\rho}} < \left(-\frac{\sigma^2}{4(\rho+\sigma-1} + 2\delta\right) \tag{58}$$

for $T > T_{R,\delta} > T_R$ and $\beta = \beta(T)$. Now, from (54)–(58), it follows that there exist points α and β, $0 < \alpha < \beta < \infty$, such that the desired inequality

$$(a[v], v)_{(\alpha,\beta)} < 0 \tag{59}$$

holds.

In the case $q(x) \leqslant 0$, $x > x_0$, and $\alpha > x_0$, we have

$$R^{1-\rho-\sigma} \int_\alpha^\beta q(x) x^\sigma \varphi^2(x)\,dx \leqslant R^{1-\rho-\sigma} \int_R^T x^\sigma q(x)\,dx$$
$$< \rho - 1 + \frac{\sigma^2}{4(\rho+\sigma-1)} - 3\delta$$

for δ sufficiently small, if $R > R_\delta$ and $T > T_R$. Using (57) and (58) we again obtain (59). Thus in the case $\rho < 1$, $\rho + \sigma < 1$, Theorem 6.8 is proved.

Next, let $\rho < 1$ and $\rho + \sigma > 1$. We start from

$$(a[v], v)_{(\alpha,\beta)} = T^{\rho+\sigma-1}\left(\frac{T^{1-\rho-\sigma} R^{\rho+\sigma-1}(1-\rho)}{1-(\alpha/R)^{1-\rho}} + \frac{\sigma^2}{4(\rho+\sigma-1)}\left\{1 - \left(\frac{R}{T}\right)^{\rho+\sigma-1}\right\}\right. \tag{60}$$
$$\left. - \frac{1-\rho}{1-(\alpha/R)^{1-\rho}} + T^{1-\rho-\sigma} \int_\alpha^\beta q(x) x^\sigma \varphi^2(x)\,dx\right).$$

Here, $\varphi^2(x)$ is now increasing on $[\alpha, R]$ if R is sufficiently close to α and decreasing on $[T, \beta]$ for every $\beta > T$. As before, but now using (51), we can determine points α and β such that

$$(a[v], v)_{(\alpha,\beta)} < 0 .$$

The case $q(x) \leqslant 0$, $x > x_0$, can also be handled as above.

If either $\rho > 1$, $\rho + \sigma < 1$, or $\rho > 1$, $\rho + \sigma > 1$, we start from (54) or (60) respectively, and again the argument is similar.

In the case $\rho = 1$, $\sigma \neq 0$, we use the test function

$$v(x) = \begin{cases} R^{\sigma/2} \log^{-1}\left(\dfrac{R}{\alpha}\right) \log\left(\dfrac{x}{\alpha}\right), & \alpha \leqslant x \leqslant R, \\ x^{\sigma/2}, \quad R \leqslant x \leqslant T, \\ T^{\sigma/2} \log^{-1}\left(\dfrac{T}{\beta}\right) \log\left(\dfrac{x}{\beta}\right), & T \leqslant x \leqslant \beta, \end{cases}$$

and obtain

$$(a[v], v)_{(\alpha,\beta)} = R^\sigma \log^{-1}\left(\frac{R}{\alpha}\right) + \frac{\sigma}{4}(T^\sigma - R^\sigma) - T^\sigma \log^{-1}\left(\frac{T}{\beta}\right) + \int_\alpha^\beta q(x) x^\sigma \varphi^2(x)\,dx \tag{61}$$

where

$$\varphi^2(x) = \begin{cases} x^{-\sigma} R^\sigma \log^{-2}\left(\frac{R}{\alpha}\right) \log^2\left(\frac{x}{\alpha}\right), & \alpha \leqslant x \leqslant R, \\ 1, R \leqslant x \leqslant T, \\ x^{-\sigma} T^\sigma \log^{-2}\left(\frac{T}{\beta}\right) \log^2\left(\frac{x}{\beta}\right), & T \leqslant x \leqslant \beta. \end{cases}$$

If $\sigma < 0$ or $\sigma > 0$ on the right-hand side of (61), the factor R^σ or T^σ respectively is taken outside a bracket and then the calculations are analogous to those for the case $\rho \neq 1$. We note that the assertions of Theorem 6.8 are also true in the case $\rho = 1$, $\sigma \neq 0$. We move on to the case $\rho + \sigma = 1$. First let $\rho \neq 1$. Using the test function (53) we obtain

$$(a[v], v)_{(\alpha,\beta)} = \frac{1-\rho}{1-(\alpha/R)^{1-\rho}} + \frac{\sigma^2}{4}\log\left(\frac{T}{R}\right) - \frac{1-\rho}{1-(\beta/T)^{1-\rho}} + \int_\alpha^\beta q(x) x^\sigma \varphi^2(x)\,dx \tag{62}$$

where in both the cases $\rho < 1$ and $\rho > 1$ $\varphi^2(x)$ is increasing on $[\alpha, R]$ and decreasing on $[T, \beta]$ for any parameters α, R, T, β, $0 < \alpha < R < T < \beta < \infty$. Therefore we can again write

$$\int_\alpha^\beta q(x) x^\sigma \varphi^2(x)\,dx = \int_{\alpha_1}^{\beta_1} x^\sigma q(x)\,dx, \quad \alpha \leqslant \alpha_1 \leqslant R, \quad T \leqslant \beta_1 \leqslant \beta. \tag{63}$$

Now on the right-hand side of (62) the factor $\log T$ is taken in front of a bracket and then the calculations are analogous to those for the case $\rho + \sigma \neq 1$.

Finally, if $\rho = 1$ $(\sigma = 0)$, then from (52) it follows that

$$\int_1^\infty q(x)\,dx = -\infty$$

and hence the corresponding assertion of Theorem 6.8 is a special case of Theorem 6.4. This completes the proof of Theorem 6.8.

In the cases $\rho = 0$, $0 \leqslant \sigma < 1$, $q(x) \leqslant 0$, $x > x_0$, or $\rho = 0$, $\sigma > 1$, $q(x) \leqslant 0$, $x > x_0$, Theorem 6.8 was proved by Nehari [45].

If we put

$$q(x) = -x^{\rho-2}\left[\tfrac{1}{4}(1-\rho)^2 + \delta(x)\right], \quad \rho \neq 1, \quad 0 < x < \infty ,$$

and $\beta = 2T$ in (62), we obtain

$$(a[v], v)_{(\alpha,\beta)} \leqslant \frac{1-\rho}{1-(\alpha/R)^{1-\rho}} + \tfrac{1}{4}(1-\rho)^2 \log\left(\frac{\beta_1}{\alpha_1}\right) - \frac{1-\rho}{1-(\beta/T)^{1-\rho}}$$

$$+ \int_{\alpha_1}^{\beta_1} x^{1-\rho} q(x)\, dx$$

$$= \frac{1-\rho}{1-(\alpha/R)^{1-\rho}} - \frac{1-\rho}{1-2^{1-\rho}} - \int_{\alpha_1}^{\beta_1} \frac{\delta(x)}{x}\, dx$$

$$= C(\alpha, R, \rho) - \int_{1}^{\beta_1} \frac{\delta(x)}{x}\, dx$$

and therefore we have the following corollary.

Corollary 6.9†

The equation

$$-(x^{\rho} y')' - x^{\rho-2}\left[\tfrac{1}{4}(1-\rho)^2 + \delta(x)\right] y = 0, \quad 0 < x < \infty, -\infty < \rho < \infty , \tag{64}$$

is oscillatory at $x = \infty$ if

$$\int_{1}^{\infty} \frac{\delta(x)}{x}\, dx = \infty . \tag{65}$$

We remark that, in the case $\delta(x) = 0$, (64) is non-oscillatory at $x = \infty$. This fact follows from the inequality

$$\int_{0}^{\infty} x^{\rho} \,|\, u' \,|^2\, dx \geqslant \tfrac{1}{4}(1-\rho)^2 \int_{0}^{\infty} x^{\rho-2} \,|\, u \,|^2\, dx, \quad u(x) \in C_0^{\infty}(0, \infty) , \tag{66}$$

† In the case $\rho = 1$, Corollary 6.9 is a special case of Theorem 6.4.

which we get immediately from the Hardy inequality (A. 21). By (66) the operator

$$A_0 u = -(x^\rho u')' - \tfrac{1}{4}(1-\rho)^2 x^{\rho-2} u, \quad D(A_0) = C_0^\infty(0, \infty),$$

is positive, so that the spectrum of the Friedrichs extension $\hat{A}$ of A_0 contains no negative point. Hence (64) is not oscillatory at $x = \infty$ if $\delta(x) = 0$. If in (64), however, $-\frac{1}{4}(1-\rho)^2$ is replaced by a smaller number ($\delta(x) = \epsilon > 0$), then (65) holds and hence (64) is oscillatory at $x = \infty$. In this sense the number $-\frac{1}{4}(1-\rho)^2$ is sharp concerning the oscillation property. In the case $\rho = 2$ this number is equal to $-\frac{1}{4}$ in accord with the fact that the spectrum of the Friedrichs extension of

$$A_0 u = -(x^2 u')' - \tfrac{1}{4} u, \quad D(A_0) = C_0^\infty(0, \infty),$$

is the half-axis $[0, \infty)$.

Finally, we also formulate oscillation criteria which refer to the point $x = 0$. We obtain these criteria by using the above test functions for the case $x = \infty$ once again.

Theorem 6.10

The equation

$$-(x^\rho y')' + q(x) y = 0, \quad 0 < x < \infty, \quad -\infty < \rho < \infty,$$

is oscillatory at $x = 0$ if one of the following conditions is satisfied.

$$1) \ \limsup_{x \to 0} x^{1-\rho-\sigma} \int_x^1 t^\sigma q(t)\, dt < \frac{\rho - 1 - |\rho - 1|}{2} + \frac{\sigma^2}{4(\rho + \sigma - 1)}, \quad \rho + \sigma < 1,$$

$$2) \ \limsup_{x \to 0} x^{1-\rho-\sigma} \int_0^x t^\sigma q(t)\, dt < \frac{1 - \rho - |1 - \rho|}{2} - \frac{\sigma^2}{4(\rho + \sigma - 1)}, \quad \rho + \sigma > 1, \tag{67}$$

$$3) \ \limsup_{x \to 0} \frac{1}{|\log x|} \int_x^1 t^{1-\rho} q(t)\, dt < \tfrac{1}{4}(1-\rho)^2.$$

If $q(x) \leqslant 0$, $0 < x < x_0$, then "lim sup" can always be replaced by "lim inf".

$$4) \ \int_0^1 \frac{\delta(x)}{x}\, dx = \infty,$$

where $q(x)$ is written as

$$q(x) = -x^{\rho-2}\left[\tfrac{1}{4}(1-\rho)^2 + \delta(x)\right] .$$

Concerning (67) for the case $q(x) \leqslant 0$, $0 < x < x_0$, $\rho = 0$, $\sigma = 2$ see also [21].

Appendix

In this appendix, as in Chapter 1, basic methods and results are given. Primarily, we prove Embedding Theorems for Sobolev spaces with weights, and we restrict ourselves to one dimension. The following notation is used.

$C^l(x_1, x_2)$, $-\infty \leqslant x_1 < x_2 \leqslant \infty$, is the set of all l-times continuously differential (complex-valued) functions on the open interval (x_1, x_2).

$C^l[x_1, x_2]$ is the set of all l-times continuously differentiable functions on the closed interval $[x_1, x_2]$, $-\infty < x_1 < x_2 < \infty$.

$C^l[x_1, x_2)$, $-\infty < x_1 < x_2 \leqslant \infty$, and $C^l(x_1, x_2]$, $-\infty \leqslant x_1 < x_2 < \infty$, are defined similarly.

$$C_0^l(x_1, x_2) = \{u(x) \mid u(x) \in C^l(x_1, x_2) \text{ and } u(x) = 0$$
$$\text{if } x_1 < x < x_u \text{ or } X_u < x < x_2\}, \quad -\infty \leqslant x_1 < x_2 \leqslant \infty .$$

$$C_0^l[x_1, x_2) = \{u(x) \mid u(x) \in C^l[x_1, x_2) \text{ and } u(x) = 0 \text{ if}$$
$$X_u < x < x_2\}, -\infty < x_1 < x_2 \leqslant \infty .$$

$C_0^l(x_1, x_2]$ is defined similarly.

$C_p^l(x_1, x_2)$, $1 \leqslant p < \infty$, consists of the functions in $C^l(x_1, x_2)$ for which the norm

$$\| u \|_{W_p^l(x_1, x_2)} = \left(\sum_{k=0}^{l} \int_{x_1}^{x_2} | u^{(k)}(x) |^p \, dx \right)^{1/p}$$

is finite.

$\dot{C}^l(x_1, x_2)$ consists of the functions in $C^l(x_1, x_2)$ for which the norm

$$\| u \|_{\dot{C}^l(x_1, x_2)} = \sum_{k=0}^{l} \sup_{x_1 < x < x_2} | u^{(k)}(x) |$$

is finite.†

† Editor's note: In the notation of R. A. Adams, *Sobolev Spaces* (Academic Press, 1975), p. 95, this space would be denoted by $C_B^l(x_1, x_2)$.

The Banach space $W_p^l(x_1, x_2)$ is obtained from $C_p^l(x_1, x_2)$ by completion in the norm $\| \cdot \|_{W_p^l(x_1,x_2)}$. It is the space of functions whose distributional derivatives up to order l belong to $L_p(x_1, x_2)$; $L_p(x_1, x_2)$ is the set of all complex-valued measurable functions defined on (x_1, x_2) for which

$$\| u \|_{L_p(x_1,x_2)} = \left(\int_{x_1}^{x_2} | u(x) |^p \, dx \right)^{1/p} < \infty, \quad 1 \leqslant p < \infty .$$

The space $W_p^l(x_1, x_2)$ is called a Sobolev space. $W_2^l(x_1, x_2)$ is a Hilbert space. The completion of $C_0^l(x_1, x_2), -\infty \leqslant x_1 < x_2 \leqslant \infty$, in the norm of the space $W_p^l(x_1, x_2)$, $1 \leqslant p < \infty$, is denoted by $\overset{\circ}{W}{}_p^l(x_1, x_2)$. $\dot{C}^l(x_1, x_2)$ is a complete space with its norm.

We also need Sobolev spaces with weights, where the weight functions are powers of x. The completion of $C_0^\infty(0, \infty)$ in the norm

$$\| u \|_{W_{\alpha,2}^l(x_1,x_2)} = \left(\sum_{k=0}^{l} \int_{x_1}^{x_2} x^{k\alpha} | u^{(k)}(x) |^2 \, dx \right)^{1/2},$$

$$-\infty < \alpha < \infty , \tag{1}$$

with $x_1 = 0$ and $x_2 = \infty$, is denoted by $W_{\alpha,2}^l(0, \infty)$. The completions of $C_0^\infty [a, \infty)$ and $C_0^\infty (0, a]$, $0 < a < \infty$, in the norm (1) with $x_1 = a$, $x_2 = \infty$ in the first case, and $x_1 = 0$, $x_2 = a$ in the second, are denoted by $W_{\alpha,2}^l [a, \infty)$ and $W_{\alpha,2}^l(0, a]$ respectively.

To prove the Embedding Theorem we require the following lemma.

Lemma A.1 [44]
Let the (complex-valued) function $w(t)$ belong to $C^l[0, \tau]$, $\tau > 0$. Then there exist positive numbers $c = c(l)$ and $\epsilon_0(l, \tau)$, independent of $w(t)$, such that

$$\int_0^\tau | w' | \, dt + \cdots + \int_0^\tau | w^{(l-1)} | \, dt \leqslant \epsilon \int_0^\tau | w^{(l)} | \, dt + c \, \epsilon^{1-l} \int_0^\tau | w | \, dt \tag{2}$$

for every ϵ with $0 < \epsilon \leqslant \epsilon_0(l, \tau)$, and $l \geqslant 2$.

Proof
First we prove (2) for $l = 2$ and a real-valued function $u(t)$. Let t_0 be a point on $[0, \tau]$ where $| u'(t) |$ is minimal. Then we have

$$0 \leqslant m = | u'(t_0) | \leqslant | u'(t) |, \quad t_0, t \in [0, \tau] .$$

Since

$$u'(t_1) = u'(t_0) + \int_{t_0}^{t_1} u''(t)\ \mathrm{d}t, \quad t_1 \in [0, \tau]\ ,$$

we obtain

$$| u'(t_1) | \leqslant | u'(t_0) | + \int_0^{\tau} | u''(t) | \,\mathrm{d}t\ ,$$

and, by integration with respect to t_1,

$$\int_0^{\tau} | u'(t) | \,\mathrm{d}t \leqslant m\tau + \tau \int_0^{\tau} | u''(t) | \,\mathrm{d}t\ .$$

If $m > 0$ $u(t)$ is either strictly increasing or strictly decreasing on $[0, \tau]$. If $u(t)$ has a zero $\bar{t}$, then

$$\int_0^{\tau} | u(t) | \,\mathrm{d}t \geqslant \int_0^{\bar{t}} m(\bar{t} - t)\,\mathrm{d}t + \int_{\bar{t}}^{\tau} m(t - \bar{t})\,\mathrm{d}t$$
$$= m[\tfrac{1}{4}\tau^2 + (\tfrac{1}{2}\tau - \bar{t})^2] \geqslant m\tfrac{1}{4}\tau^2\ .$$

If $u(t)$ has no zero on $[0, \tau]$, then obviously we have

$$\int_0^{\tau} | u(t) | \,\mathrm{d}t \geqslant m\tfrac{1}{2}\tau^2\ .$$

Therefore in both cases we obtain

$$m\tau \leqslant 4\tau^{-1} \int_0^{\tau} | u(t) | \,\mathrm{d}t$$

and hence

$$\int_0^{\tau} | u'(t) | \,\mathrm{d}t \leqslant 4\tau^{-1} \int_0^{\tau} | u(t) | \,\mathrm{d}t + \tau \int_0^{\tau} | u''(t) | \,\mathrm{d}t\ . \tag{3}$$

If $m = 0$, (3) is again true. Obviously we also have

$$\int_{t_1}^{t_1+\epsilon} | u'(t) | \,\mathrm{d}t \leqslant 4\epsilon^{-1} \int_{t_1}^{t_1+\epsilon} | u(t) | \,\mathrm{d}t + \epsilon \int_{t_1}^{t_1+\epsilon} | u''(t) | \,\mathrm{d}t\ ,$$
$$0 \leqslant t_1 < t_1 + \epsilon \leqslant \tau\ . \tag{4}$$

If now the interval $[0, \tau]$ is divided into subintervals, the longest of which has length ϵ with $0 < \epsilon \leqslant \frac{1}{2}\tau$ and the shortest has length $\frac{1}{2}\epsilon$ at least, then from (4) we obtain

$$\int_0^\tau |u'(t)|\,dt \leqslant 8\epsilon^{-1}\int_0^\tau |u(t)|\,dt + \epsilon\int_0^\tau |u''(t)|\,dt, \quad 0 < \epsilon \leqslant \tfrac{1}{2}\tau. \tag{5}$$

This proves (2) in the case $l = 2$ and $u(t)$ is real-valued. Now let $u(t) = \text{Re } w(t)$ and $v(t) = \text{Im } w(t)$. Then, from (5) for $u(t)$ and $v(t)$ and since

$$(a^2 + b^2)^{1/2} \leqslant a + b \leqslant \sqrt{2}\,(a^2 + b^2)^{1/2}, \quad a \geqslant 0, \quad b \geqslant 0,$$

it follows by addition that

$$\int_0^\tau |w'(t)|\,dt \leqslant 8\sqrt{2}\epsilon^{-1}\int_0^\tau |w(t)|\,dt + \sqrt{2}\epsilon\int_0^\tau |w''(t)|\,dt.$$

Writing ϵ in place of $\sqrt{2}\epsilon$ we have

$$\int_0^\tau |w'(t)|\,dt \leqslant 16\epsilon^{-1}\int_0^\tau |w(t)|\,dt + \epsilon\int_0^\tau |w''(t)|\,dt, \quad 0 < \epsilon \leqslant \tau/\sqrt{2}. \tag{6}$$

To prove (2) for $l > 2$, we first use an induction argument to establish the existence of positive constants a_n and b_n such that

$$\sum_{\nu=1}^{n-1}\int_0^1 |w^{(\nu)}(t)|\,dt \leqslant a_n\int_0^1 |w(t)|\,dt + b_n\int_0^1 |w^{(n)}(t)|\,dt. \tag{7}$$

Assuming (7), then, and applying (7) to $w'(t)$, we also have

$$\sum_{\nu=1}^{n}\int_0^1 |w^{(\nu)}(t)|\,dt \leqslant (a_n + 1)\int_0^1 |w'(t)|\,dt + b_n\int_0^1 |w^{(n+1)}(t)|\,dt$$

from which, because of (6) ($\tau = 1$), we obtain

$$\begin{aligned}\sum_{\nu=1}^{n}\int_0^1 &|w^{(\nu)}(t)|\,dt \\ &\leqslant 16(a_n + 1)\epsilon^{-1}\int_0^1 |w(t)|\,dt + \epsilon(a_n + 1)\int_0^1 |w''(t)|\,dt \\ &+ b_n\int_0^1 |w^{(n+1)}(t)|\,dt.\end{aligned}$$

If we choose ϵ so small that $\epsilon(a_n+1)<\frac{1}{2}$ and take the middle term on the right-hand side of the last inequality over to the left, we obtain an estimate of the form

$$\sum_{\nu=1}^{n}\int_0^1 |w^{(\nu)}(t)|\,dt \leqslant a_{n+1}\int_0^1 |w(t)|\,dt + b_{n+1}\int_0^1 |w^{(n+1)}(t)|\,dt\,,$$

and (7) now clearly follows by induction on n.

Making the transformation $s=\epsilon t$, $w(t)=w_1(s)$, in (7), we have

$$\sum_{\nu=1}^{n-1}\epsilon^{\nu-1}\int_0^{\epsilon} |w_1{}^{(\nu)}(s)|\,ds \leqslant a_n\epsilon^{-1}\int_0^{\epsilon} |w_1(s)|\,ds + \epsilon^{n-1}b_n\,.$$

If this inequality is multiplied by ϵ^{2-n} with $0<\epsilon<1$, we obtain

$$\sum_{\nu=1}^{n-1}\int_0^{\epsilon} |w_1{}^{(\nu)}(s)|\,ds \leqslant a_n'\epsilon^{1-n}\int_0^{\epsilon} |w_1(s)|\,ds + \epsilon b_n'\int_0^{\epsilon} |w_1{}^{(n)}(s)|\,ds. \tag{8}$$

If we write t and w in place of s and w_1 respectively and consider the interval $[0,\tau]$ as a union of intervals with lengths between $\frac{1}{2}\epsilon$ and ϵ, where $0<\epsilon\leqslant\min(1,\frac{1}{2}\tau)$, then by addition of the corresponding inequalities of type (8) we obtain

$$\sum_{\nu=1}^{n-1}\int_0^{\tau} |w^{(\nu)}|\,dt \leqslant a_n''\epsilon^{1-n}\int_0^{\tau} |w|\,dt + \epsilon b_n''\int_0^{\tau} |w^{(n)}|\,dt,$$

$$0<\epsilon<\epsilon_0(n,\tau)\,.$$

This proves Lemma A.1 for $l=n$, if $\epsilon b_n''$ is replaced by ϵ.

Theorem A.2

If $l>k\geqslant 0$ and $1<p<\infty$, then the space $W_p^l(x_1,x_2)$, $-\infty\leqslant x_1<x_2\leqslant\infty$, is continuously embedded into the space $\dot{C}^k(x_1,x_2)$† and, for every $\epsilon>0$, there is a constant C_ϵ such that

$$\|u\|_{\dot{C}^k(x_1,x_2)} \leqslant \epsilon\,\|u^{(l)}\|_{L_p(x_1,x_2)} + C_\epsilon\,\|u\|_{L_p(x_1,x_2)} \tag{9}$$

for all $u(x)\in W_p^l(x_1,x_2)$. If (x_1,x_2) is a finite interval, the embedding from $W_p^l(x_1,x_2)$ into $\dot{C}^k(x_1,x_2)$ is compact, and the functions of $W_p^l(x_1,x_2)$ can be continued to $[x_1,x_2]$ in such a way that they belong to $C^k[x_1,x_2]$.

† That is, each function in $W_p^l(x_1,x_2)$ is equal to a function in $\dot{C}^k(x_1,x_2)$ almost everywhere.

Proof

It is sufficient to prove (9) for functions in $C_p^l(x_1, x_2)$. Then (9) follows by completion of $C_p^l(x_1, x_2)$ in the W_p^l-norm. We start with the obvious equation

$$u(x) = u(t) + (x-t)u'(t) + \cdots + \frac{(x-t)^{l-1}}{(l-1)!}\, u^{(l-1)}(t) + \frac{1}{(l-1)!}\int_t^x (x-s)^{l-1} u^{(l)}(s)\,\mathrm{d}s\,. \tag{10}$$

On integrating with respect to t over an interval ω, of length $|\omega|$, containing x and lying in (x_1, x_2), we obtain

$$|u(x)| \leqslant \sum_{k=0}^{l} |\omega|^{k-1} \int_\omega |u^{(k)}(t)|\,\mathrm{d}t, \quad x \in \omega\,. \tag{11}$$

In (11), we use (2), with $\epsilon = 1$, to obtain

$$|u(x)| \leqslant (1 + |\omega|^{l-1}) \int_\omega |u^{(l)}(t)|\,\mathrm{d}t + (C_1 + |\omega|^{-1}) \int_\omega |u(t)|\,\mathrm{d}t\,, \tag{12}$$

and then by the Hölder inequality, we have

$$|u(x)| = (1 + |\omega|^{l-1})\,|\omega|^{1/q} \left(\int_\omega |u^{(l)}|^p\,\mathrm{d}t\right)^{1/p} + (C_1 + |\omega|^{-1})\,|\omega|^{1/q} \left(\int_\omega |u|^p\,\mathrm{d}t\right)^{1/p}$$

$$\leqslant (1 + |\omega|^{l-1})\,|\omega|^{1/q}\, \|u^{(l)}\|_{L_p(x_1,x_2)} + (C_1 + |\omega|^{-1})\,|\omega|^{1/q}\, \|u\|_{L_p(x_1,x_2)}\,, \tag{13}$$

where $p^{-1} + q^{-1} = 1$. Then (13) implies (9) for the case $k = 0$ if $|\omega|$ is made to depend on ϵ suitably. If, in (12), $u(x)$ is replaced by $u^{(k)}(x)$ and l by $I - k$ and then (2) is used again, we obtain (9) for general k, $0 \leqslant k < l$.

We now prove the compactness of the embedding when (x_1, x_2) is a finite interval. In order to use the theorem by Arzelà, we show that the property

$$\| u \|_{W_p^l(x_1,x_2)} \leqslant C, \quad p > 1 , \tag{14}$$

implies the equi-continuity of the functions $u^{(k)}(x)$, $0 \leqslant k < l$, on (x_1, x_2). In (10) we replace x and t by $x + h$ and x respectively to obtain

$$u(x+h) - u(x) = hu'(x) + \cdots + \frac{h^{l-1}}{(l-1)!} u^{(l-1)}(x) + \frac{1}{(l-1)!} \int_x^{x+h} (x+h-s)^{l-1} u^{(l)}(s) \, ds . \tag{15}$$

Then, by the Hölder inequality and (9) and (14), we have

$$| u(x+h) - u(x) |^p \leqslant C_1 \left(| h |^p \, | u'(x) |^p + \cdots + | h |^{p(l-1)} \, | u^{(l-1)}(x) |^p + | h |^{p(l-1)+p/q} \int_{x_1}^{x_2} | u^{(l)} |^p \, ds \right) \leqslant C_2 (| h |^p + \cdots + | h |^{p(l-1)} + | h |^{p(l-1)+p/q}) . \tag{16}$$

From (16) it follows that the functions $u(x)$ are equi-continuous because the right-hand side of (16) tends to zero when $h \to 0$.

Now, in (16), we replace $u(x)$ by $u^{(k)}(x)$ and l by $l - k$ to obtain

$$\| u(x+h) - u(x) \|_{\dot{C}^k(x_1,x_2,h)} < \epsilon, \quad | h | < \delta(\epsilon) , \tag{17}$$

where

$$(x_1, x_2, h) = (x_1, x_2) \cap (x_1 - h, x_2 - h) ,$$

from which we have the compactness of the embedding when $1 \leqslant k \leqslant l - 1$. Since, by (17), each function $u(x) \in W_p^l(x_1, x_2)$ and its derivatives up to order $l - 1$ are uniformly continuous on (x_1, x_2), $u(x)$ possesses limits at the boundary points x_1 and x_2 and can be so continued on $[x_1, x_2]$ that after the continuation $u(x)$ belongs to $C^{l-1}[x_1, x_2]$. Theorem A.2 is now proved.

The standard method used here to prove the Embedding Theorem has been extended in [44] to prove the Sobolev Embedding Theorems for arbitrary dimension N.

We go on to consider weighted spaces.

Theorem A.3

1) Let $u(x) \in W^n_{\alpha,2}(0, \infty)$. Then for every $\epsilon > 0$ there is a C_ϵ such that

$$\int_0^\infty x^{k\alpha} | u^{(k)} |^2 \, dx \leqslant \epsilon \int_0^\infty x^{n\alpha} | u^{(n)} |^2 \, dx + C_\epsilon \int_0^\infty | u |^2 \, dx , \tag{18}$$

$$1 \leqslant k \leqslant n - 1, \quad -\infty < \alpha < \infty .$$

2) Let $a > 0$, $\alpha \leqslant 2$ and $u(x) \in W^n_{\alpha,2}\,[a, \infty)$. Then for every $\epsilon > 0$ there exists a $C_{a,\epsilon}$ such that

$$\int_a^\infty x^{k\alpha} | u^{(k)} |^2 \, dx \leqslant \epsilon \int_a^\infty x^{n\alpha} | u^{(n)} |^2 \, dx + C_{a,\epsilon} \int_a^\infty | u |^2 \, dx,$$

$$1 \leqslant k \leqslant n - 1 . \tag{19}$$

Here, a general $C_{a,\epsilon} = C_\epsilon$ can be chosen if $a \geqslant a_0 > 0$.

3) Let $a > 0$, $\alpha \geqslant 2$ and $u(x) \in W^n_{\alpha,2}(0, a]$. Then for every $\epsilon > 0$ there exists a $C_{a,\epsilon}$ such that

$$\int_0^a x^{k\alpha} | u^{(k)} |^2 \, dx \leqslant \epsilon \int_0^a x^{n\alpha} | u^{(n)} |^2 \, dx + C_{a,\epsilon} \int_0^a | u |^2 \, dx,$$

$$1 \leqslant k \leqslant n - 1 , \tag{20}$$

and, if $0 < a \leqslant a_0 < \infty$, a general $C_{a,\epsilon} = C_\epsilon$ can be chosen again.

Proof

First we prove (18) for $u \in C_0^\infty(0, \infty)$. In the following estimates the Hardy inequality

$$\int_0^\infty x^\beta | u |^2 \, dx \leqslant \frac{4}{(\beta + 1)^2} \int_0^\infty x^{\beta+2} | u' |^2 \, dx, \quad \beta \neq -1, \quad u(x) \in C_0^\infty(0, \infty) , \tag{21}$$

is used and this can be proved by integration by parts as follows. If $u(x)$ is real-valued, we have

$$\int_0^\infty x^\beta u^2 \, dx = -2(\beta + 1)^{-1} \int_0^\infty u u' t^{\beta+1} \, dt$$

$$\leqslant 2 \, | \beta + 1 |^{-1} \left(\int_0^\infty x^\beta u^2 \, dx \right)^{1/2} \left(\int_0^\infty x^{\beta+2} (u')^2 \, dx \right)^{1/2} .$$

Squaring this inequality we obtain (21) for a real-valued $u(x)$. For complex-valued $u(x)$, (21) follows immediately on considering the real and imaginary parts separately.

We prove (18) by induction on n. Let $n = 2$, $k = 1$ and, first, $\alpha \neq \frac{1}{2}$. We suppose that $u(x)$ is real-valued. Using integration by parts, the Schwarz inequality, and the inequality

$$ab \leqslant \epsilon a^2 + \tfrac{1}{4}\epsilon^{-1}b^2 ,$$

where a and b are real numbers and $\epsilon > 0$, we obtain

$$\int_0^\infty x^\alpha u u' \, dx = (-\alpha) \int_0^\infty x^{\alpha-1} u' u \, dx - \int_0^\infty x^\alpha u'' u \, dx$$

$$\leqslant |\alpha| \left(\int_0^\infty x^{2\alpha-2} (u')^2 \, dx \right)^{1/2} \left(\int_0^\infty u^2 \, dx \right)^{1/2}$$

$$+ \left(\int_0^\infty x^{2\alpha} (u'')^2 \, dx \right)^{1/2} \left(\int_0^\infty u^2 \, dx \right)^{1/2}$$

$$\leqslant \epsilon_1 \int_0^\infty x^{2\alpha-2} (u')^2 \, dx + C_{\epsilon_1} \int_0^\infty u^2 \, dx + \epsilon_2 \int_0^\infty x^{2\alpha} (u'')^2 \, dx + C_{\epsilon_2} \int_0^\infty u^2 \, dx .$$

Together with (21) this gives

$$\int_0^\infty x^\alpha (u')^2 \, dx \leqslant [4\epsilon_1 (2\alpha - 1)^{-2} + \epsilon_2] \int_0^\infty x^{2\alpha} (u'')^2 \, dx + (C_{\epsilon_1} + C_{\epsilon_2}) \int_0^\infty u^2 \, dx$$

$$= \epsilon \int_0^\infty x^{2\alpha} (u'')^2 \, dx + C_\epsilon \int_0^\infty u^2 \, dx ,$$

which is (18) in the case under consideration. If $\alpha = \frac{1}{2}$, we argue as follows:

$$\int_0^\infty x^{1/2} u' u' \, dx = \tfrac{1}{4} \int_0^\infty x^{-1/2} (u^2)' \, dx - \int_0^\infty x^{1/2} u'' u \, dx$$

$$= -\tfrac{1}{8} \int_0^\infty x^{-3/2} u^2 \, dx - \int_0^\infty x^{1/2} u'' u \, dx \leqslant \epsilon \int_0^\infty x (u'')^2 \, dx + C_\epsilon \int_0^\infty u^2 \, dx ,$$

and (18) is now proved for $n = 2$.

Continuing the induction argument, we now assume that (18) is true for real-valued functions $u(x) \in C_0^\infty(0, \infty)$, a number $n \geqslant 2$, and all $k = 1, \ldots, n-1$. We begin with $\alpha \neq (n+1)^{-1}$. Then

$$\int_0^\infty x^{n\alpha} u^{(n)} u^{(n)} \, dx$$

$$= (-n\alpha) \int_0^\infty x^{n\alpha-1} u^{(n)} u^{(n-1)} \, dx - \int_0^\infty x^{n\alpha} u^{(n+1)} u^{(n-1)} \, dx$$

$$\leqslant | n\alpha | \left(\int_0^\infty x^{\alpha(n+1)-2} (u^{(n)})^2 \, dx \right)^{1/2} \left(\int_0^\infty x^{\alpha(n-1)} (u^{(n-1)})^2 \, dx \right)^{1/2}$$

$$+ \left(\int_0^\infty x^{\alpha(n+1)} (u^{(n+1)})^2 \, dx \right)^{1/2} \left(\int_0^\infty x^{\alpha(n-1)} (u^{(n-1)})^2 \, dx \right)^{1/2}$$

$$\leqslant \epsilon_1 \int_0^\infty x^{\alpha(n+1)-2} (u^{(n)})^2 \, dx + C_{\epsilon_1} \int_0^\infty x^{\alpha(n-1)} (u^{(n-1)})^2 \, dx$$

$$+ \epsilon_2 \int_0^\infty x^{\alpha(n+1)} (u^{(n+1)})^2 \, dx + C_{\epsilon_2} \int_0^\infty x^{\alpha(n-1)} (u^{(n-1)})^2 \, dx .$$

From this and (21), we obtain

$$\int_0^\infty x^{n\alpha} (u^{(n)})^2 \, dx$$

$$\leqslant (4\epsilon_1 [\alpha(n+1) - 1]^{-2} + \epsilon_2) \int_0^\infty x^{\alpha(n+1)} (u^{(n+1)})^2 \, dx$$

$$+ (C_{\epsilon_1} + C_{\epsilon_2}) \int_0^\infty x^{\alpha(n-1)} (u^{(n-1)})^2 \, dx$$

$$= \epsilon \int_0^\infty x^{\alpha(n+1)} (u^{(n+1)})^2 \, dx + C_\epsilon \int_0^\infty x^{\alpha(n-1)} (u^{(n-1)})^2 \, dx ,$$

and hence, by (18),

$$\int_0^\infty x^{n\alpha}(u^{(n)})^2\,dx$$

$$\leqslant \epsilon \int_0^\infty x^{\alpha(n+1)}(u^{(n+1)})^2\,dx + C_\epsilon \tilde{\epsilon} \int_0^\infty x^{\alpha n}(u^{(n)})^2\,dx + C_\epsilon\, C_{\tilde{\epsilon}} \int_0^\infty u^2\,dx\,.$$

If ϵ is fixed and $\tilde{\epsilon}$ is chosen so small that $C_\epsilon \tilde{\epsilon} \leqslant \frac{1}{2}$, this gives

$$\frac{1}{2}\int_0^\infty x^{n\alpha}(u^{(n)})^2\,dx \leqslant \epsilon \int_0^\infty x^{\alpha(n+1)}(u^{(n+1)})^2\,dx + C_\epsilon\, C_{\tilde{\epsilon}} \int_0^\infty u^2\,dx\,.$$

Replacing ϵ by $\frac{1}{2}\epsilon$, we have therefore

$$\int_0^\infty x^{n\alpha}(u^{(n)})^2\,dx \leqslant \epsilon \int_0^\infty x^{\alpha(n+1)}(u^{(n+1)})^2\,dx + C_\epsilon \int_0^\infty u^2\,dx\,, \tag{22}$$

which is (18) with $n+1$ in place of n and $k = n$. If $\alpha = (n+1)^{-1}$, we obtain

$$\int_0^\infty x^{n\alpha}(u^{(n)})^2\,dx \leqslant \epsilon \int_0^\infty x^{\alpha(n+1)}(u^{(n+1)})^2\,dx + C_\epsilon \int_0^\infty x^{\alpha(n-1)}(u^{(n-1)})^2\,dx$$

as above in the case $\alpha = \frac{1}{2}$. Then (22) follows again, by (18) for $k = n-1$. Thus (18) holds for $k = n$ and $n+1$ in place of n. Then we obtain (18) for $1 \leqslant k \leqslant n-1$ from (22) by noting that

$$\int_0^\infty x^{k\alpha}(u^{(k)})^2\,dx \leqslant \epsilon \int_0^\infty x^{n\alpha}(u^{(n)})^2\,dx + C_\epsilon \int_0^\infty u^2\,dx$$

$$\leqslant \epsilon\tilde{\epsilon} \int_0^\infty x^{\alpha(n+1)}(u^{(n+1)})^2\,dx + (\epsilon C_{\tilde{\epsilon}} + C_\epsilon) \int_0^\infty u^2\,dx\,.$$

The induction argument is now complete and (18) is proved for real-valued $u(x) \in C_0^\infty(0,\infty)$ and all $n = 1, 2, \ldots$. The result for complex-valued $u(x)$ follows immediately. Finally we may take $u(x) \in W^n_{\alpha,2}(0,\infty)$ in (18), because such a function can be approximated by C_0^∞-functions in the norm $\|\cdot\|_{W^n_{2}{}^{\alpha}(0,\infty)}$.

We prove (19) first for $a = 1$. By Theorem A.2 there is a $C > 0$ such that

$$\int_0^1 |u^{(k)}|^2\,dx \leqslant C\left(\int_0^1 |u^{(n)}|^2\,dx + \int_0^1 |u|^2\,dx\right),$$

$$u \in C^n[0,1],\quad 0 \leqslant k \leqslant n\,. \tag{23}$$

In (23) the transformation $x = \omega^{-1}\xi$, $\omega > 0$,† is made, then ξ is replaced by x, and after that $u(\omega^{-1}x)$ by $u(x)$. This gives

$$\int_0^\omega |u^{(k)}|^2\,dx \leqslant C\,\omega^{2(n-k)} \int_0^\omega |u^{(n)}|^2\,dx + C\,\omega^{-2k} \int_0^\omega |u|^2\,dx,$$

$$u \in C^n[0, \omega]\,. \qquad (24)$$

If, in (24), ω is bounded by $0 < \omega < \frac{1}{2}$ and the interval $[0, 1]$ is covered by intervals the lengths of which are chosen between $\frac{1}{2}\omega$ and ω, then addition of the corresponding inequalities (24) gives

$$\int_0^1 |u^{(k)}|^2\,dx \leqslant C\,\omega^{2(n-k)} \int_0^1 |u^{(n)}|^2\,dx + C\,\omega^{-2k} \int_0^1 |u|^2\,dx\,,$$

$$0 < \omega < \tfrac{1}{2}, \quad u \in C^n[0, 1]\,,$$

with a new constant C. Clearly we also have

$$\int_1^2 |u^{(k)}|^2\,dx \leqslant C\,\omega^{2(n-k)} \int_1^2 |u^{(n)}|^2\,dx + C\,\omega^{-2k} \int_1^2 |u|^2\,dx$$

from which we obtain immediately

$$\int_1^2 x^{k\alpha} |u^{(k)}|^2\,dx \leqslant C\,\omega^{2(n-k)} \int_1^2 x^{n\alpha} |u^{(n)}|^2\,dx + C\,\omega^{-2k} \int_1^2 |u|^2\,dx,$$

$$0 < \omega < \tfrac{1}{2}, \quad u \in C^n[1, 2]\,. \qquad (25)$$

By means of the transformations $x = 2^{-\nu}\xi$, $\nu = 0, 1, 2, \ldots$, it follows from (25) that

$$\int_{2^\nu}^{2^{\nu+1}} x^{k\alpha} |u^{(k)}|^2\,dx$$

$$\leqslant C 2^{(n-k)(2-\alpha)\nu} \omega^{2(n-k)} \int_{2^\nu}^{2^{\nu+1}} x^{n\alpha} |u^{(n)}|^2\,dx + C 2^{-k(2-\alpha)\nu} \omega^{-2k} \int_{2^\nu}^{2^{\nu+1}} |u|^2\,dx\,, \qquad (26)$$

†Editor's note: it is convenient not to maintain the notation $|\omega|$ for the length of an interval here.

and, if we now choose $\omega = \omega_\nu$ for different ν in such a way that

$$\omega_\nu 2^{(1-\alpha/2)\nu} = \epsilon, \quad 0 < \epsilon < \tfrac{1}{2}, \quad \nu = 0, 1, 2, \ldots,$$

then, by addition in (26), we obtain

$$\int_1^\infty x^{k\alpha} \mid u^{(k)} \mid^2 dx \leqslant C\, \epsilon^{2(n-k)} \int_1^\infty x^{n\alpha} \mid u^{(n)} \mid^2 dx$$

$$+ C\, \epsilon^{-2k} \int_1^\infty \mid u \mid^2 dx, \quad 0 < \epsilon < \tfrac{1}{2}, \quad u \in C_0^n [1, \infty).$$

Finally, the transformation $x = a^{-1}\xi$ gives

$$\int_a^\infty x^{k\alpha} \mid u^{(k)} \mid^2 dx \leqslant C\, \epsilon^{2(n-k)} a^{(2-\alpha)(n-k)} \int_a^\infty x^{n\alpha} \mid u^{(n)} \mid^2 dx$$

$$+ C\, \epsilon^{-2k} a^{-k(2-\alpha)} \int_a^\infty \mid u \mid^2 dx, \quad u \in C_0^n [a, \infty).$$

Replacing $C\, \epsilon^{2(n-k)} a^{(2-\alpha)(n-k)}$ by ϵ again, we obtain (19) for $u \in C_0^n [a, \infty)$. Then (19) also follows for $u \in W_{\alpha,2}^n [a, \infty)$.

To prove (20) we start from (25). Clearly we also have

$$\int_{1/2}^1 x^{k\alpha} \mid u^{(k)} \mid^2 dx \leqslant C\, \omega^{2(n-k} \int_{1/2}^1 x^{n\alpha} \mid u^{(n)} \mid^2 dx + C\, \omega^{-2k} \int_{1/2}^1 \mid u \mid^2 dx,$$

$$0 < \omega < \omega_0, \quad u \in C^n [\tfrac{1}{2}, 1], \quad 1 \leqslant k \leqslant n-1. \quad (27)$$

The transformations $x = 2^{-\nu}\xi$, $\nu = 0, -1, -2, \ldots$, now give

$$\int_{2^{\nu-1}}^{2^\nu} x^{k\alpha} \mid u^{(k)} \mid^2 dx$$

$$\leqslant C\, 2^{(n-k)(2-\alpha)\nu} \omega^{2(n-k)} \int_{2^{\nu-1}}^{2^\nu} x^{n\alpha} \mid u^{(n)} \mid^2 dx + C\, 2^{-k(2-\alpha)\nu} \omega^{-2k} \int_{2^{\nu-1}}^{2^\nu} \mid u \mid^2 dx.$$

In this inequality we put $\omega = \epsilon 2^{(\alpha/2-1)\nu}$, $0 < \epsilon < \omega_0$, to obtain

$$\int_{2^{\nu-1}}^{2^\nu} x^{k\alpha} |u^{(k)}|^2 \, dx \leqslant C\,\epsilon^{2(n-k)} \int_{2^{\nu-1}}^{2^\nu} x^{n\alpha} |u^{(n)}|^2 \, dx + C\,\epsilon^{-2k} \int_{2^{\nu-1}}^{2^\nu} |u|^2 \, dx . \tag{28}$$

Then addition of the equations (28) for $\nu = 0, -1, -2, \ldots$, gives

$$\int_0^1 x^{k\alpha} |u^{(k)}|^2 \, dx \leqslant C\,\epsilon^{2(n-k)} \int_0^1 x^{n\alpha} |u^{(n)}|^2 \, dx$$

$$+ C\,\epsilon^{-2k} \int_0^1 |u|^2 \, dx, \quad 0 < \epsilon < \omega_0, \quad u \in C_0^n(0, 1] ,$$

and, by the transformation $x = a^{-1}\xi$, $a > 0$, we obtain

$$\int_0^a x^{k\alpha} |u^{(k)}|^2 \, dx \leqslant C\,\epsilon^{2(n-k)} a^{(2-\alpha)(n-k)} \int_0^a x^{n\alpha} |u^{(n)}|^2 \, dx$$

$$+ C\,\epsilon^{-2k} a^{-k(2-\alpha)} \int_0^a |u|^2 \, dx . \tag{29}$$

Replacing $C\,\epsilon^{2(n-k)} a^{(2-\alpha)(n-k)}$ by ϵ again, we obtain from (29)

$$\int_0^a x^{k\alpha} |u^{(k)}|^2 \, dx \leqslant \epsilon \int_0^a x^{n\alpha} |u^{(n)}|^2 \, dx + C_{a,\epsilon} \int_0^a |u|^2 \, dx, \tag{30}$$

$$u \in C_0^n(0, a], \quad 1 \leqslant k \leqslant n-1 ,$$

and (30) also holds for $u \in W^n_{\alpha,2}(0, a]$. This proves (20) and completes the proof of Theorem A.3.

In conclusion, we describe the method of smoothing of functions. The function

$$\varphi(x) = \begin{cases} \gamma \exp[(x^2 - 1)^{-1}], & |x| < 1, \quad \gamma > 0 \\ 0, & |x| \geqslant 1 , \end{cases}$$

belongs to $C_0^\infty(-\infty, \infty)$. Let γ be chosen so that $\| \varphi \|_{L_1(-\infty,\infty)} = 1$. By means of $\varphi(x)$ a function $u(x) \in L_{1,\text{loc}}(x_1, x_2)$ can be smoothed in terms of the function $u_\epsilon(x)$ defined by

$$u_\epsilon(x) = \int_{|y| \leq 1} u(x - \epsilon y)\varphi(y)\, dy, \quad \epsilon > 0, \quad x_1 + \epsilon < x < x_2 - \epsilon .^{\dagger} \tag{31}$$

By a simple transformation, we can write (31) as

$$u_\epsilon(x) = \epsilon^{-1} \int_{|y-x| \leq \epsilon} u(y)\varphi\left(\frac{x-y}{\epsilon}\right) dy, \quad x_1 + \epsilon < x < x_2 - \epsilon . \tag{32}$$

Here, ϵ is called the radius of smoothing. The properties of $u_\epsilon(x)$ are stated in the following theorem.

Theorem A.4

1) If $u(x) \in L_{1,\text{loc}}(x_1, x_2)$, then $u_\epsilon(x) \in C^\infty(x_1 + \epsilon, x_2 - \epsilon)$ for every $\epsilon > 0$.

2) Let $u(x) \in C^l(x_1, x_2)$ and let $\epsilon_0 > 0$ be arbitrary. Then each derivative

$$u_\epsilon^{(k)}(x), \quad 0 \leq k \leq l, \quad 0 < \epsilon \leq \epsilon_0 ,$$

tends uniformly to the corresponding derivative $u^{(k)}(x)$ on $(x_1 + \epsilon_0, x_2 - \epsilon_0)$ when $\epsilon \to 0$.

3) If $u(x)$ belongs to $W_2^l(x_1, x_2)$ and $\epsilon_0 > 0$ is arbitrary, then the smoothed function $u_\epsilon(x)$ tends to $u(x)$ in the $\| \cdot \|_{W_2^l}$-norm on $(x_1 + \epsilon_0, x_2 - \epsilon_0)$ when $\epsilon \to 0$.

Proof
See [52].

†Editor's note: The terms mollification and regularization are also applied to $u_\epsilon(x)$.

References

[1] Achiezer, N. I. and Glazman, I. M. (1961). *Theory of Linear Operators in Hilbert Space*, Ungar, New York.

[2] Balslev, E. (1966). The singular spectrum of elliptic differential operators in $L^p\ (R_n)$, *Math. Scand.*, **19**, 193-210.

[3] Birman, M. S. (1959). Perturbation of quadratic forms and the spectrum of singular boundary value problems (Russian), *DAN SSSR*, **125**, 471-474.

[4] Birman, M. S. (1964). Transcript of lecture.

[5] Borg, G. (1951). On the point spectra of $y'' + (\lambda - q(x))\, y = 0$, *Amer. J. Math.*, **73**, 122-126.

[6] Brinck, I. (1959). Self-adjointness and spectra of Sturm-Liouville operators, *Math Scand.*, **7**, 219-239.

[7] Coppel, W. A. (1965). *Stability and asymptotic behaviour of differential equations*, D. C. Heath, Boston.

[8] Dunford, N. and Schwartz, J. T. (1963). *Linear Operators*, Part II, Interscience.

[9] Eastham, M. S. P. (1970). The least limit point of the spectrum associated with singular differential operators, *Proc. Camb. Phil. Soc.*, **67**, 277-281.

[10] Eastham, M. S. P. (1973). *The Spectral Theory of Periodic Differential Equations*, Scottish Academic Press.

[11] Eastham, M. S. P. (1976). On the absence of square-integrable solutions of the Sturm-Liouville equation, *Proceedings of the Dundee Conference on Differential Equations. Lecture Notes in Mathematics*, **564**, Springer.

[12] Eastham, M. S. P. (1977). Sturm-Liouville equations and purely continuous spectra, *Niew Archief v. Wiskunde* (3), **25**, 169-181.

[13] Everitt, W. N. (1972). On the spectrum of a second-order linear differential equation with a p-integrable coefficient, *Applicable Analysis*, **2**, 143-160.

[14] Friedrichs, K. O. (1948). Criteria for the discrete character of the spectra of ordinary differential operators, *Studies and Essays presented to R. Courant.* Interscience.

[15] Glazman, I. M. (1965). *Direct Methods of Qualitative Spectral Analysis of Singular Differential Operators*, I.P.S.T. Jerusalem.

[16] Hartman, P. and Wintner, A. (1948). Criteria of non-degeneracy for the wave equation, *Amer. J. Math.*, **70**, 295-308.

[17] Hille, E. (1948). Nonoscillation theorems, *Trans. Amer. Math. Soc.*, **54**, 234-252.

[18] Hinton, D. (1970). Continuous spectra of second-order differential operators, *Pacific J. Math.*, **33**, 641-643.

[19] Hinton, D. (1974). Continuous spectra of an even order differential operator, *Illinois J. Math.* **18**, 444-450.

[20] Hinton, D. B. and Lewis, R. T. (1975). Discrete spectra criteria for singular differential operators with middle terms, *Math. Proc. Camb. Phil. Soc.*, **77**, 337-347.

[21] Hinton, D. B. and Lewis, R. T. (1978). Oscillation theory at a finite singularity, *J. Diff. Equations,* **30**.

[22] Hunt, R. W, and Namboordiri, M. S. T. (1970). Solution behaviour for general self-adjoint differential equations, *Proc. London Math. Soc.*, **21**, 637-650.

[23] Ismagilov, R. S. (1961). Conditions for the semi-boundedness and discreteness of the spectrum of one-dimensional differential operators (Russian), *DAN SSSR,* **140**, 33-36.

[24] Kamke, E. (1950). Differentialgleichungen reeller Funktionen, Leipzig.

[25] Kato, T. (1952). Note on the least eigenvalue of the Hill equation, *Quart. Appl. Math.*, **10**, 292-294.

[26] Kato, T. (1966). *Perturbation Theory of Linear Operators,* Springer.

[27] Kneser, A. (1893). Untersuchungen über die reellen Nullstellen der Integrale linearer Differentialgleichungen, *Math. Ann.*, **42**, 409-435.

[28] Kreith, K. (1963). Differential operators with a purely continuous spectrum, *Proc. Amer. Math. Soc.*, **14**, 809-811.

[29] Kreith, K. (1973). *Oscillation Theory,* Lecture Notes in Mathematics No. **324**, Springer.

[30] Leighton, W. (1952). On self-adjoint differential equations of second order, *J. London Math. Soc.*, **27**, 37-47.

[31] Leighton, W. and Nehari, Z. (1958). On the oscillation of self-adjoint linear differential equations of the fourth order, *Trans. Amer. Math. Soc.*, **89**, 325-377.

[32] Lewis, R. T. (1974). The discreteness of the spectrum of self-adjoint, even order, one term, differential operators, *Proc. Amer. Math. Soc.*, **42**, 408-482.

[33] Lewis, R. T. (1974). Oscillation and nonoscillation criteria for some self-adjoint even order linear differential operators, *Pacific J. Math.* **51**, 221-234.

[34] Lewis, R. T. (1976). The existence of conjugate points for self-adjoint differential equations of even order, *Proc. Amer. Math. Soc.*, **56**, 162-166.

[35] Molčanov, A. M. (1953). The conditions for the discreteness of the spec-

trum of self-adjoint second-order differential equations (Russian), *Trudy Moskov Mat. Obsc.*, **2**, 169–200.

[36] Müller-Pfeiffer, E. (1970). Spektraleigenschaften eindimensionaler Differentialoperatoren höherer Ordnung, *Studia Math.*, **34**, 183–196.

[37] Müller-Pfeiffer, E. (1971). Notwendige un hinreichende Bedingungen für die Diskretheit des Spektrums Sturm-Liouvillescher Operatoren, *Acta Math. Acad. Sci. Hung.*, **22**, 385–392.

[38] Müller-Pfeiffer, E. (1974). Relativ vollstetige Störungen von gewöhnlichen Differentialoperatoren höhere Ordnung, *Studia Math.*, **49**, 153–160.

[39] Müller-Pfeiffer, E. (1974). Gewöhnliche Differentialoperatoren mit rein stetigem Spektrum, *Math. Nachr.*, **62**, 163–178.

[40] Müller-Pfeiffer, E. (1975). Ergänzung zur Arbeit "Gewöhonliche Differentialoperatoren mit rein stetigem Spectrum", *Math. Nachr.*, **67**, 255-263.

[41] Müller-Pfeiffer, E. (1976), Sturm-Liouvillesche Operatoren mit stetigem Spektrum auf Intervallen (Λ, ∞), $-\infty \leqslant \Lambda < \infty$, *Math. Nachr.*, **72**, 7–13.

[42] Müller-Pfeiffer, E. (1977). Über den Einfluss der Randbedingungen auf die Existenz von Eigenwerten bei gewöhnlichen Differentialoperatoren höherer Ordnung, *Math. Nachr.*, **76**, 123–137.

[43] Müller-Pfeiffer, E. (1978). Eine Abschätzung für das Infimum des Spektrums des Hillschen Differentialoperators, *Publ. Math. Debrecen*, **25**, 35–40.

[44] Müller-Pfeiffer, E and Weber, A. (1974). Ein Beweis der Sobolevschen Einbettungssätze, *Math. Nachr.* **64**, 277–288.

[45] Nehari, Z. (1957). Oscillation criteria for second-order linear differential equations, *Trans. Amer. Nath. Soc.*, **85**, 428–445.

[46] Naimark, M. A. (1968). *Linear Differential Operators*, Harrap.

[47] Putnam, C. R. (1955). Integrable potentials and half-line spectra, *Proc. Amer. Math. Soc.*, **6**, 243–246.

[48] Riesz, F. and Sz.-Nagy, B. (1955). *Lectures on Functional Analysis*, Ungar, New York.

[49] Schechter, M. (1972). *Spectra of Partial Differential Operators*, North Holland.

[50] Schechter, M. (1973). On the essential spectrum of partial differential operators bounded from below. *Scripta Math.*, **29**, 5–16.

[51] Sears, D. B. (1951). On the spectrum of a certain differential equation, *J. London Math. Soc.*, **26**, 205–210.

[52] Sobolev, S. L. (1963). Applications of functional analysis in mathematical physics, *Amer. Math. Soc. Translations*, **7**.

[53] Swanson, C. A. (1968). *Comparison and Oscillation Theory of Linear Differential Equations*, New York and London.

[54] Titchmarsh, E. C. (1962). *Eigenfunctions Expansions Associated with Second-order Differential Equations*, Part I, 2nd ed., Oxford.

[55] Triebel, H. (1965). *Vorlesungen über Funktionalanalysis.*

[56] Triebel, H. (1972). *Höhere Analysis,* Berlin.
[57] Wallach, S. (1948). On the location of spectra of differential equations, *Amer. J. Math.,* **70**, 833–841.
[58] Walter, J. (1972). Absolute continuity of the essential spectrum of $-d^2/d^2 + q(t)$ without monotony of q, *Math. Z.,* **129**, 83–94.
[59] Weidman, J. (1967). Zur Spektraltheorie von Sturm-Liouville Operatoren, *Math Z.,* **98**, 268–302.
[60] Wintner, A. (1949). A criterion of oscillatory stability, *Quart. Appl. Math.,* **7**, 115–117.

Index